G   C

"IN THE MIDWES[T] [THERE IS ALWAYS] THE THREAT THAT WHAT SEEMS SO NORMAL, SO CALM AND PEACEFUL—A LITTLE CLOUD, FOR INSTANCE, OR A REFRESHING BREEZE OR A LIGHT RAIN SHOWER—CAN SUDDENLY TURN UGLY, VIOLENT AND LETHAL."

—from *Final Harvest*

"The IN COLD BLOOD of the 1980s . . . details the shattering of an American dream . . . a stunning social document . . . a tapestry of faded dreams and the demise of long-held attitudes . . . lifts the veil from the essence and mores of small-town life . . . hones in on lives that symbolize millions."

—*American Way*

"GRAPHIC AND MOVING . . . shows violence boiling up and over."

—*Chicago Tribune*

"WELL-PACED, SUSPENSEFUL TRUE CRIME."
—*Washington Post Book World*

Andrew H. Malcolm is the Chicago Bureau Chief of *The New York Times*. Born in Cleveland, raised in rural Ohio, and educated in Indiana and Illinois, he has for four decades personally witnessed the painful changes sweeping the nation's industrial and agricultural heartland. An award-winning correspondent, Mr. Malcolm has also reported for *The Times* from San Francisco, New York, Japan, Korea, and Indochina.

FINAL HARVEST

An American Tragedy

Andrew H. Malcolm

A SIGNET BOOK

NEW AMERICAN LIBRARY

Grateful acknowledgment is made to Ellen C. Masters for permission to
reprint an excerpt from ''Gordon Halica'' from *The New Spoon River*,
by Edgar Lee Masters, Collier Books, 1968, p. 46.

FOR CONNIE M.

ACKNOWLEDGMENTS

I could never have successfully completed a project of this magnitude without the cooperation—large and small, time-consuming and fleeting—of scores of people too numerous to list comprehensively here. First, I should like to thank many colleagues at *The New York Times*, especially A. M. Rosenthal, Seymour Topping, Arthur Gelb, and David R. Jones for their longtime professional encouragement and guidance, understanding, and, perhaps most of all, for their friendship. Also Paul Delaney, Irv Horowitz, and Martha Miles, and the panoply of professional editors whose deft touches, known and unknown to me, improved my stories on this and other subjects and who fought, often successfully, to get the stories as much space as I was so certain they deserved. I have also benefited from the assistance and company of Wayne King and Mike Elliott, Sara Hennings, Patrick Marx, Paul Levy, and Steve Olson; from the knowledge and experience of David Ostendorf, Dan Levitas, and Peter Zevenbergen; the insights of C. Conway Smith; the love of words and confidence of Arthur G. Hughes; the kind counseling of Eugene Kennedy even amid his own deadlines; the guidance and encouragement of Jonathan Segal and Julian Bach; and the broad exposure to my own land and native region provided over many years in several small and large towns by my parents, Ralph and Beatrice Malcolm. I must acknowledge the generous donation of time and thought and patience given me by the understanding people of Ruthton and Ivanhoe, Minnesota, and Brownwood and Paducah, Texas, during my research stints in their communities and lives. I must also add here a special note of thanks to Susan Blythe for helping to trace, without conditions, the life of her family and for reliving in intimate detail for me and my notebooks the most painful times of her life. And to Lynnette Thulin and Karen Rider and the many, many other friends, relatives, acquaintances, and investigators of the protagonists on all sides of this American tragedy, I owe a debt of gratitude for sharing their memories, thoughts, hopes, and trust, and for always answering my many questions, even at strange hours and even if I cannot list their names here. These would include a special friend whose expertise helped me wade through and decipher many intricacies of the legal portion of this drama that would have remained arcane and lost to general readers including myself. I also want to thank Robert A. Berg and Michael J. O'Gorman for all of their insights and fellowship. I will always value the insights, patience, and company of our children, Christopher, Spencer, and Emily. And I could never express fully how deeply my appreciation runs to my wife, Connie, whose priceless and timely encouragements, helpful writing suggestions, and sensitive human and professional insights may yet lead me to comprehend—someday—the strange inner workings of an editor's mind.

A. H. M.

CONTENTS

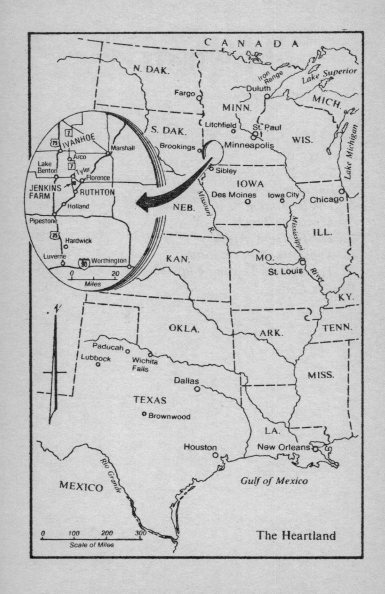

The Heartland

"Didja ever notice that prey have their eyes on the side of their head and predators have their eyes on the front?

"Now, you look where man's eyes are."

—CHARLES SNOW,
retired soldier

FOREWORD

At 9:10 A.M. Central Daylight Time on September 29, 1983, I was on the telephone from Chicago with my boss, David R. Jones, national news editor of *The New York Times*. After less than a year back in the United States following the better part of a decade abroad, I had been struck by some major social, economic, and structural changes underway all across the nation's heartland, a broad, diverse section where I had grown up and worked, in what was fast becoming a bygone time.

I had completed a lengthy series of articles for *The Times* on these changes two months before that rainy Thursday morning. And in the phone call we agreed that I should return to the subject, especially the mounting financial difficulties being encountered by farmers, families, and businessmen across the region and the impact of this far beyond. It was an important subject, we agreed before hanging up, that threatened possible violence and old ways of living.

At that precise moment about five hundred miles away, four men—three native midwesterners and an immigrant from the East, each with his own dream for the future—were acting out a deadly drama in the fog of an overgrown patch of prairie.

That same morning, after a hurried phone call, I found myself on an airplane flying toward those fertile fields of southwestern Minnesota, the most rural, farming-oriented congressional district in the nation, and preparing to encounter a group of people and towns I had never seen before but knew so well, and would come to know even better in the months ahead.

That fatal farmyard confrontation, the manhunt, and the resulting trial came to symbolize many of the deep changes happening slowly in so many communities important to the country's soul. Such violence would be repeated and continue in the seasons that followed. As recently as last month, it made headlines (*IOWA FARMER SLAYS THREE, SELF*) and sadness, but, I fear, not widespread understanding of the underlying changes haunting the Heartland. As a journalist for two decades, I have often been told about fascinating stories that speak to larger issues. Often these pitches are true, though rarely do they fully live up to advance billing. However, the story of Rudy Blythe and James Jenkins, the striking parallels in their lives, and the dramatic details of their confrontation and the ensuing destruction far exceeded my expectations or dreams as a professional storyteller, even after two years of reconstructing this sad tale.

I have tried to recount the many human stories in this saga as fairly and accurately as possible, relying on documentation, firsthand accounts obtained through scores of interviews, published reports, personal memories, my own notes and observations, and court transcripts. I have discovered over the years that there is no single, all-powerful truth. This work is a detailed compilation of many truths, a measurement of a nation's change taken at one moment, hopefully providing a sense of perspective and scale to an otherwise fleeting local story so easily lost in the swift-flowing stream of louder events that seem so important, for a day anyway.

—*ANDREW H. MALCOLM*
Chicago, January 14, 1986

STORM WATCH

It was such a happy time.

On September 28, 1983, Susan Blythe returned from a long trip to Texas with a car and a trailer full of her family's furniture and clothing. To her waiting husband, Rudy, it seemed she was finally accepting their move away from their friends in the fine homes of north Dallas back to their modest home in an almost modest town in rural southwestern Minnesota. He knew how Susan felt about their smalltown life, and how he felt about it. Ever since they had moved to the Midwest from Philadelphia, Rudy had wanted to own a small country bank. It was a dream that had dominated his thoughts, his savings, his career plans, even his weekends, when he would bundle his wife and son into the car and cruise the streets of the rural Midwest looking for likely bank candidates to buy. The attraction was the lack of bureaucracy. Being his own boss. A feeling of power. The sense of community. Doing things when he wanted, the way he wanted. The intimacy with clients. And, not least, the intoxicating sense, for once, of being needed that would surely earn him wide acceptance among the men he always sought out. It all appealed to him so much that after the final papers were signed on Thanksgiving in 1977, he marched Susan and little Rolph outside in the snow and cold wind and had his father take a picture of the three of them by the bank's brick walls beneath the stylish sign, Buffalo Ridge State Bank. Finally, it was his.

The two males—big, tall, heavyset Rudy, a good bit overweight for his forty-two years from his nervous eating of the last few months, and short, stocky Rolph, an instant guard on any football team of eleven-year-olds—had organized a family party for Susan's thirty-ninth birthday that homecoming night, complete with a store-bought white cake that said, "Happy Birthday, Susan." Cake was not on Rudy's diet, but this was a special night, much more than a normal birthday. He'd have baked one himself had he known how or had the patience to learn. These months of trouble at the bank and living alone, with Susan and Rolph still in Texas, were about to end. The dream was back on track. So he had bought the cake at the grocery and had it specially decorated.

He and Rolph were anxiously but happily waiting at home as Susan drove slowly through the darkened, leaf-strewn streets of the little town of Ruthton, some five hundred miles northwest of Chicago, that Wednesday night. She knew the community well from the four years she had spent in its confines before the move to Texas. Word spreads quickly about everything in a small town, especially when it concerns someone as important as The Banker. All 332 residents knew Susan Blythe was coming back.

Shortly after 10:00 P.M. she turned off the main highway onto Leo Street and pulled up in front of her house on the wrong side of the road. That doesn't matter in places like Ruthton where everybody thinks they know everything about everybody. Lyle Landgren, the deputy sheriff, would likely cruise by on his nightly rounds later, but he wouldn't do anything. He knew the house and the car and the family situation. And they knew his; the Blythes had loaned him the money for his own home. As Susan pulled up, she saw some movement in the front window. Rolph's chubby face ducked behind the lighted curtains. When the front door opened, there was an explosion of hugs and kisses as the pent-up homemade warmth swept into the cool autumn night. All the presents were piled by the hearth, including several from Susan's parents back in New York. Her parents' gifts were clothes—matching skirt, sweater, and blazer. "All mothers give clothes at birthdays," said Susan. And the three of them laughed at the universal truth—even Susan, who gave clothes at birthdays. Rudy gave his wife a sweater and skirt and a nightgown—no necklace this time. Rolph handed her a box containing a new pair of slippers. Rudy, who had already passed what they called "The Big Four Oh," gave Susan a card that read "Perhaps you're over the hill." Inside her husband had written, "But the valley is not deep."

The big man and the little man sang "Happy Birthday" then. And they lit a few candles on the cake; they were men, but when it came to Mom, they knew better than to light all thirty-nine. The cake was cut, the rainbow sherbet dished out, and they all sat down at the table.

It had taken Susan two long days to make the trip from Texas, even with her friend Sharon Fadness along. They talked and talked while Susan drove and drove for hour after hour as if her life was set on slow motion. Usually Susan felt a sense of urgency in her life—make a plan, concisely, efficiently, compromise here and there if necessary, and then get on with it. Sometimes when things felt as slow as they did during those endless, droning hours in her northbound car, she was reminded of that chilly, idyllic afternoon in the park outside Philadelphia when she and Rudy first dated not many years before. A huge field was surrounded by woods. "Pretend the other side of the field is freedom," said Rudy impulsively, "and we're going to run to freedom." They had crept to the edge of the clearing, crouched, and looked for border guards. Then they had made a mad dash into the open, holding hands, running across the uneven ground, slipping and stumbling and laughing out loud and then reminding each other with "Shsssh" that the unseen enemy somewhere in the trees might spot their movements and swing his evil black rifle toward them. And so they ran harder and harder. It always took far too long to reach safety, Susan thought. Whenever she recalled that silliness, her voice grew warm and she felt close to Rudy. And he said he remembered it fondly too.

After her safe arrival at home for the birthday party, it seemed that everything was moving so rapidly, like a sled rushing downhill toward the woods. There was audible joy to this high-speed homecoming celebration. It was exciting. They all were talking so quickly. They grabbed for the time together as if it were running out. For Susan, life was moving with such velocity that it seemed out of control . . . and she wanted to slow it down to savor more.

The sherbet was melting, and it was past Rolph's bedtime. But everybody that night wanted to talk at once: Sue about the trip to Texas and their friends there and how the trailer had started to fall apart; Rolph about school and football; and Rudy about the bank. He had a laundry list of things he urgently wanted to tell her, and every time the subject turned away from the bank, he would remember some other piece of financial news. "Oh, Dad," Rolph said at one point, "all you ever talk about is the bank."

"He's right," said Susan. "Can't this wait until later?"

"Oh," her husband continued, barging on and pretending to almost forget the big news, "the DeRuyter loan was paid off." Susan did listen to that—ninety thousand dollars was no small sum. As a bank board member, she knew about that loan. It was one of those that had been bothering Rudy, considering the continuing agricultural recession and the declining value of farmland and collateral. But her husband patted the pocket of his blue oxford shirt. He pulled out a pink deposit slip. And he smiled.

Then Rolph was tucked into bed. No time for Dad to read to him tonight. That was okay. Everybody was back together again, and not just for a short visit. Susan and Rudy strolled outside into the chilly, wet September night; no doubt about it, winter was coming to town. Lyle cruised by in his patrol car and flashed his lights and waved; too late to beep in Ruthton. The thought of unpacking everything that night in the light rain was too much for the Blythes. "Leave it," said Rudy, "and I'll help you tomorrow after closing."

"Okay," said Susan. "Watch this." She ran off to the car to show her husband her newfound skill at maneuvering a tricky trailer. After she backed up across the lawn several times, Susan gave up. She pulled the car into the garage, and they laughed together as they went indoors.

"Oh, I forgot to tell you," Rudy said later as his mind continued down the mental list of news items, "I have an appointment tomorrow to show the house, the old Jenkins farm." Susan needed no reminding about that place. For four years they had owned that farm, ever since that son of a bitch sold his mortgaged cattle, declared bankruptcy, and stuck them with a thirty-thousand-dollar loss.

"Who wants to buy it?" Susan said through the bathroom door as she finished her shower.

Her husband replied. But the water was running. It sounded like one of those Scandinavian names.

"Who?" said Susan, happy that they might finally be unloading the farm with the haunted house just north of town but too tired then to really care. She opened the bathroom door, the steam billowing into the cooler bedroom. Her husband was in mid-sentence.

". . . don't know him. No one seems to. He's from up north somewhere. I tried to check. But he phoned and I'll meet him out at the farm tomorrow morning."

"Well, will you meet him before nine?" asked Susan. " 'Cause I could take Rolph to school in the bank car and get it back in time for your appointment." She knew how strongly Rudy felt against using the company car for personal business, but the windshield wipers didn't work on their own car. It was eighteen miles to the Pipestone school, and just three blocks to the bank.

When they crawled into bed, it was eleven-thirty. Too late for Rudy to read any more of *American Caesar*. The clock radio was on softly. Susan had tuned it from Rudy's favorite classical music station to the easy listening one in Sioux Falls. "I can't go to sleep to Wagner," she always said. The last thing they heard as they drifted off to sleep, together again, was the weather forecast for Thursday. Storms were coming.

The morning dawned cold and drizzly and foggy. Hard to wake up. Hard to drive. Hard to harvest. Hard to see except for bright colors. The tired family was running late. Susan felt harried. She quickly whipped up some toast, poured the juice for Rolph, and dumped some hot water in a cup with some instant coffee from a little sample jar.

"How long do you have to be married," Rudy said as he took a sip and made a face, "to get a real cup of coffee in the morning?"

"You know how to make it yourself," she said. "Get off your keister and do it. I've got to get going." It was a simple, eminently forgettable remark, the kind of crack, when said with a smile, that seems to require no apology. Within hours, those nineteen words would sound harsher in Susan's mind than she ever intended.

Susan was still in the bedroom when Rudy got up from the breakfast table, hugged his son, and donned his bright yellow rain slicker, his birthday present of a few months before. "Will you be back by nine with the car?" he called. And Susan said yes, but she didn't hear him leave for the bank downtown. He was in such a hurry, in fact, that he did not kiss his wife good-bye. So when she emerged from the bedroom all fresh and dressed in her exercise clothes and talking to her husband, Rudy was gone. She didn't think much about it then, for she was in her own hurry.

At the bank that morning there was a minor legal problem. Deems "Toby" Thulin, Rudy's new loan officer, was a little embarrassed. He had been out driving the previous night in a pickup truck the bank had recently repossessed, and was stopped by the state highway patrol. The truck's license plate sticker had expired, and the officer issued Toby a ticket. No problem, said Rudy. In a few minutes they'd just drive the six miles up County Road 7 to the Tyler Town Hall. They'd get the new registration in the works, get Toby back to work reviewing the bank's problem loans, and Rudy could make his ten o'clock appointment out at the Jenkins place.

Rudy liked Toby, who was even newer in Ruthton than his boss. Toby was thirty-seven, a Vietnam veteran, his résumé said, wiry, athletic, liked a cold beer with the guys on a hot Sunday after a couple of softball games. He had bumped around the Midwest for a number of years working for banks and loan companies and doing some odd construction work. But things were not really expanding in his world, especially for someone without a college education. Toby was laid off or the boss couldn't afford or wasn't disposed to dispense the benefits he had promised at hiring. Toby was not succeeding financially. He had tried to hide it all; men are supposed to in the country. But his wife, Lynnette, saw the impact in the longer, quieter periods: his quicker temper, his frequent absences, sometimes even a wavering of his religious faith. She got part-time work and made and sold crafts from their basement to bring in a few extra dollars for them and their three little girls. Lynnette had even put out a feeler for her husband with an executive search company. There were interviews, nothing for many months, and then along came a real possibility—a job with a little bank in Ruthton, Minnesota. The opening with the Blythes looked good, although soon after Toby began work, the strains of their situation had prompted the Thulins into a trial separation. They had a marriage counseling session set for the next weekend.

Getting the truck registration papers didn't take long. Rudy, who rarely carried any cash, simply wrote out a check to the state for $37.68. The two bankers were chatting as usual over coffee in the cafe. It was hunting season, and there was nothing Toby liked more than hunting or fishing, or talking about hunting or fishing. He was a very good shot, and among friends he might make some quiet reference to all the shooting he'd seen and heard and done over in Vietnam. Rudy liked hunting and fishing too, or rather he liked the fraternity; the guys out in the boat or on the dock, the late dinner telling stories around the fire. It was soon nearly nine o'clock and time for Rudy to take Toby back to the bank in Ruthton, get a couple of things done, and then go back out to the Jenkins place for the appointment at ten.

In the closing days of September the fields along County Road 7, like those along thousands of miles of prairie roads from Ohio to Colorado and from Minnesota south to Texas, are lined with dead or dying things. Green grasses turning yellow. Waves of leaves scuttling back and forth. Tumbleweeds piled against fences. The dried, crunched carcasses of birds or rabbits or a slow-moving possum, the likely victims of a pickup truck's wheels, lying in the dirt, picked clean by brazen crows with beady eyes and oily feathers. Some of the corn has been picked by then. Much has not, and in the winds, the crisp, crumpled cornstalks rub against one another across the empty fields in a strange kind of dry death rattle.

On the drive past these fields back to Ruthton, Rudy, the perfectionist, decided to stop by the abandoned Jenkins farm just to check so that everything would go smoothly with the potential buyer an hour later. Rudy did not believe in locking up the empty houses and barns he was selling. He figured if anybody was going to break into an isolated farmhouse, a locked door would not stop him. And an unlocked door might save a lot of damage. Anyway, there wasn't anything to take out of the old house. When Jenkins had left four years ago, he'd even torn out the bathtub.

It was raining again as Rudy approached the Jenkins farm. The station wagon's windows were closed, so Rudy and Toby didn't hear their slowing tires swishing on the wet pavement. But other men did. When the car turned into the gravel driveway, these other men also heard the crunching of stones. They ran and hid. In a few minutes Rudy and Toby would be running, too.

In Texas at that hour Charles Snow was in his tiny office with all the military battle prints on the wall. After thirty-one years of military service, the blunt, tough-talking maintenance supervisor had retired from uniform but retained the taut mind and manner of the top sergeant he had been for so long. It was just another day at work for Snow, who could live on his army pension but liked the camaraderie of men.

Snow hoped to get away a little early that day for some work out on his small counry ranch. He likes the open air, the immense country sky free of wires and walls, printed rules and confining clocks, and he is drawn to those who feel the same. Inside his cubicle of an office, Snow is easily irritated, often harried and distracted. He looks at someone or something while his mind is somewhere else and then swears loudly, at precisely what, no one is quite sure. But no one wants to ask either. Walking through his school maintenance garage into the liberating outdoors, Snow shrugs his shoulders as if shucking off something invisible. Then he becomes calm, talkative, and thoughtful. Snow had no idea then, of course, of the deadly drama unfolding some 1,700 miles to the north. But it would leave him late in the afternoon in his pickup truck, asking himself, out loud, some difficult questions.

Two hundred miles away Sheriff Frank Taylor, the only lawman on duty in the only town of substance in sandy Cottle County, Texas, would soon be getting into his patrol car to check on a couple of distant ranches. A dedicated Dallas football fan, he was anticipating Sunday's game on television; the Cowboys were playing the Vikings. Sheriff Taylor had never been to Minnesota, but he knew one thing about that goddamned place from the television: it was always colder than a witch's tit up there. And those Minnesota guys played rough. It would be a dramatic encounter. Good thing the game was on a Sunday afternoon. Saturday evening and he'd be out all night handling the drunks and the wrecks and the shootings. But Sunday afternoon is always nice and peaceful-like in Paducah, Texas.

First thing Thursday morning, however, the sheriff thought he might wander by the sandy little cemetery on the southeast side of town, the Garden of Memories. It is a quiet place, like most rural cemeteries not yet surrounded by suburban shopping malls. It is a suitable setting for memories, near town to avoid the long horseback ride of yesteryear yet far enough to be out of sight and out of mind at times. Sheriff Taylor went out there sometimes when things were slow and he wanted to think for a spell with his older brother, Mickey. They had been very close, but one day not too long before, Sheriff Taylor had gotten a phone call. His brother had been found on a country lane shot under the arm through the heart with a deer rifle, another rural suicide without warning. Sheriff Taylor had seen violent death up close; in fact, he had caused a few in his line of work. That couldn't be helped. But he would never forget the sight of all Mickey's thick, dark blood drying in a huge, ugly red splotch in the sand.

Rudolph H. Blythe, Sr., had gone to the country club not far from the retirement home he and his wife, Dorothy, had in Gainesville, Florida. He spent the morning talking with friends in his investment club. His wife would stay at home again with the cleaning woman; Dot's legs, stricken with polio, were bothering her again and walking had become difficult. To pick up her spirits, perhaps they'd call their son Rudy tonight and see if Susan got home all right from Texas.

Bill Slater should have retired too, but the old railroad man couldn't keep away from the trains. He was out that morning helping New York State crews survey the repair work on a crumbling old rail line near Utica. His wife, Alice, was at home watching a TV talk show and thinking she really ought to be cleaning the kitchen.

Robert Berg, an agent for the Bureau of Criminal Apprehension, Minnesota's state detective force, was enjoying a slow, sleepy morning and a hot cup of black coffee in his office in Worthington, Minnesota, near the Iowa border. He had just finished up a case. Someone had beaten a two-year-old girl to death, punched her so hard in the stomach her arteries ruptured. Berg had been called out on the homicide at 4:00 A.M., on Father's Day, and had spent the better part of three months investigating the case.

Tom Fabel, the tall, gangly, and eloquent deputy attorney general of Minnesota, was in his St. Paul office. He'd come downtown early to do as much legal work as possible before the phones started ringing; the father of three little girls, he hoped to get away by midmorning for the family's long drive north to their country cabin and an extended weekend in the woods.

Mike O'Gorman, another BCA agent, was a hundred miles south of St. Paul investigating a "house shooting." An angry husband, seeking reconciliation with his estranged wife, had stormed out of his farmhouse and then apparently shot a .22-caliber bullet through the residence, narrowly missing the woman. Guns are common in the countryside. In fact, many rural homes have more guns than people. Investigating this near murder would consume O'Gorman's day and evening, and keep him away from a radio and the news that would take control of his family's life for many months.

Judge Walter H. Mann was in his chambers doing paperwork. Karen Rider was at home humming in the shower. Lynnette Thulin was in Professor Hilton's classroom, when she felt slightly nauseous. She looked at the clock—9:10. At that moment Abe Thompson, sheriff of Lincoln County in southwestern Minnesota, glanced at his watch. Suddenly both his phone lines lit up simultaneously.

That was strange.

2

PARADISE LOST

I sat and looked at the river
Riffled and stirred by the wind;
But I saw that the depths of the river
Were moved by the under stream.
　　　　　　　—EDGAR LEE MASTERS,
　　　　　　　　　The New Spoon River

On the surface, Ruthton, Minnesota, looks like any of
a thousand other midwestern towns, an aging collec-
tion of small houses scattered on straight streets around
a downtown one or two blocks long. Maybe there is a
village square with a county courthouse in the middle
or at least a little-used band shell. It is all plunked
down on the prairie for precise reasons now lost or
forgotten. It probably had a lot to do with some bearded
fellow from the East who was not a farmer, although
he had to deal with them, but who owned a little
parcel of land and carried a gold watch in his vest
pocket with a fuzzy photograph of his wife inside the
shiny cover. Ruthton's founder's name was W. H.
Sherman. He owned the townsite. His wife's name
was, of course, Ruth.

Perhaps the railroad went through town. Maybe it
still does, although only freights do, and by now the
old depot will have been abandoned or hauled away to
become a boutique or a garage somewhere. A river or
creek may flow by, though no one above the age of
three would ever think of drinking that liquid anymore. Maybe two roads still come together there, and
where there is an intersection in the Midwest, there is

almost always a general store that was once a gas station, too; today, the dials on its abandoned pumps stare blindly out at passersby, who can't believe the old prices frozen in time. Now the new general store is called a "convenience store." It's cheese, once home-made in giant blocks by nearby farmers who traded them for store goods, comes now in heat-sealed, one-portion, handi, color-printed plasti-paks from a far-away factory that has its products delivered by the soda pop company. A computer prints the invoice.

There will likely be a tall, rusting water tower, usually silver, in these towns. The town's name will be painted in black on two sides, and scrawled around the high walkway, about teen-ager-tall, will be the year of a high school graduating class or two and the names of a couple of particularly bold boys—for that one night—and their girl friends, who waited, giggling, in the car so far below, holding the half-empty bottles of Coca-Cola.

The streets that had broad old elm trees will have broad old elm stumps now and much more sun, which means trouble for the pachysandra that every street's old widow once nourished in the shade up the block. By the time the new disease-resistant strains of trees are tall and thick and broad enough to cast noticeable, blessed shadows on August afternoons, the old lady will be long gone. And a red metal swing set will stand in the old garden with a stark bald spot on the ground just beneath the swing seat where passing little feet scuff the toes of shoes they never have to polish with their own hands. The widow might recognize her old garden. But she'd shake her head at how families with several cars and televisions, with wives with jobs outside the home, don't seem to care about the same things anymore. It doesn't take long, she'd think, for plants and life to go to seed.

There will be a bicycle or two nearby, safely cast aside in yards, to be found there days later in precisely the same place. No one thinks a thing about that. There will be doors locked only once a year when everyone must search for the key to give to a neighbor for safekeeping during vacation. There may or may not be stop signs at the street corners. There likely won't be street signs; people generally refer to the lanes by lasting human landmarks, as in, "He lives over by Darrell's house."

The downtown will be quite, both blocks of it. A few cars will come and go all day, with a brief flurry around ten for the morning coffee break, a few more at twelve for the noon meal, and a smaller gathering for afternoon coffee. There should be a car or two down by the lone surviving market and the post office with its squeaking door and yellowing "Wanted" notices, the occasional whine of an air wrench from the Standard Oil station, and the quiet grinding of ancient starters on muddy pickup trucks parked by the curb, where no one would ever think of cementing parking meters because they might chase the few remaining customers out to a shopping mall. In the country, where the air is so clean you can't even see it, many pickups will carry an air freshener, a green gizmo that hangs on the mirror and emits the scent of pine trees artificially created by chemicals in a distant city factory.

Ruthton's downtown is a quiet place and folks pretty much like it that way. Besides Jensen's Food Market, there are storefronts for a run-down laundromat, an American Legion office, Duane DeBettignies's Buffalo Ridge Printing and Gazette Shop, a New-Used Furniture and Upholstery store that looks mainly used, and a tavern with a Schlitz beer light but no name sign because everyone knows it is the Polar Bar. Ruthton's bar also has no phone, preventing anxious spouses from interrupting the "happy hour" when all the stories told along the counter are sad. The row of storefronts includes a Senior Citizens Center, where no one ever bothers to change the "Open" sign even after midnight, a "Plbg and Htg" office, a tiny library, a few abandoned buildings, a Tom Sawyer-style whitewashed fence on one vacant lot. Next to Alene's restaurant and bowling alley sits a small park with fading framed tributes to high school sports achievements. These naturally include the 1975 football team that beat Balaton 35–12 to go 9–0 for the season and the 1981 basketball team whose 17–0 record and state tournament appearance got Ruthton and Rudy Blythe so excited that practically the entire town traveled to the final game and the bank's fancy hospitality suite. The nicest building, of course, is the bank whose big windows and newer bricks shine across the street from the restaurant where Gary Lindahl's mom serves up her homemade pies. Most of the town's homes along the curbless streets are aging, with a little garage out back, its white paint peeling in spots. There is the odd car abandoned in its owner's front yard as a rusting eyesore that heals over time. Some corners have lone streetlights that shine their wan beams down on the little Girl Scouts and Cub Scouts en route home by six. By 9:00 P.M. the streetlights are the brightest things visible, save for the cafe's ubiquitous red Coca-Cola machine, whose familiar *ker-thumk* dispensing sound is sometimes the only sign of late-night life. Occasionally, however, a blast of noise and warm, stale air gushes out the opening bar door as someone stumbles through to try and drive home.

The fields are never far from such towns. They sit there like time itself, silent and dark and taken for granted. In winter the fields are sleeping. In spring they are wet and foul. By fall they are dried and tired. But come summer the fields are alive with animals and insects and microbes and tractors and lush green growth that is lovingly tended by men in dusty baseball caps who swarm about the edges and dash across the middle with implements that put finishing touches on the soil as a sculptor might before his clay dries.

Each morning a thin band of pink light eases gently into the eastern sky just over the tasseled tops of the mile-long cornfields. As if on a signal, the crickets' chirping gives way to the birds' singing. The stars start to fade. The breezes die down. The swarming insects retreat from their frantic winged worship of the yard lights. And soon the sun is back out, baking the corn and the soybeans and anyone or anything not in the shade. It is a wonderful time of year—life unfolding as it should, rich in its promise of harvests to come, rewarding of its bounty for long labors lavished, reassuring in its regularity.

The work is hard, very hard in the spring: preparing the fields, discing them, plowing them, planting them, feeding them, applying medicine, and waiting patiently for the sun and soil, the rain and God to work their magic. In the barn they invest the same care in the animals, the hogs, the milk cows and the cattle. Set the alarm for three each spring morning to check the cows for birthing problems, especially the first-time mothers. Watch them eat more grain the day before a winter storm, as if they knew, which they do. Be alert for those little mood swings, droopy ears or tail, dull eyes, that presage a natural storm of disease. Help the neighbors 'cause, sure as shootin', they'll help you someday. Soon those little hogs are big hogs. The cattle are filling out nicely. And those little green shoots of corn, 20,000 or more to the acre, will start reaching for the sun. If everything goes right, in 126 days those 20,000 little shoots should become 14,000,000 kernels, bulging yellow nuggets of nutrition that should pay for all the seeds and chemicals of spring plus perhaps a little for the labor too.

There will be some rough times, for sure. I remember back in the 1930s, everyone's father would start out. All the children's eyes would turn to their mother, pleading silently. But she would say nothing. Then everyone's eyes would glaze, except for the wife, who would stop spooning out the potatoes to remain attentive to the memory, though she knew it well by heart and experience, too. It might be the sawdust story, how that useless substance sold for more than wheat. Or the one about grandpa having to sell off all the animals or the bank taking them for some reason. Or the yechy tale about all the grasshoppers—that got everyone's attention again. They came hopping in from somewhere by the millions to consume everything of value before them, and then, worse yet, to jump around in everyone's hair, like this. The little girls would scream, and the little boys, wishing they could too, would settle for a face reflecting distaste.

But the story endings were always the same. Hard work will cure anything. When I was little, the father would say, I had one pair of shoes, my brother's old ones. We milked before school. No television in those days, no sirree! And there wasn't but one schoolroom. But my daddy worked hard. And he taught me to work hard. And I worked hard. And now look what you've got—television, telephone, your own radio. Look at all that machinery out there. And the food on this table. Hard work. That's the secret. You don't put it into the fields, you don't get it out. Hard, honest work, he'd say, predictably slowing for emphasis. And . . . don't . . . you . . . forget . . . it, . . . son. Then, as if to rub in the lesson, he'd tousle the child's hair with his huge hand, which was the loving punctuation to the serious message.

Because there was so much to do, there was an organization to daily life, a semiassigned list of chores, except in the kitchen, where Mom ruled most times unless a grandma was over. Up around dawn, two hours' work outdoors, a hearty breakfast, more long hours outdoors, a hearty dinner with everyone appearing from different fields promptly at noon. More long hours outdoors, a lighter supper after five, when rural roads suddenly became deserted, some more hours of work outdoors, followed likely by ice cream. Then, by nine or ten, the sheds were shut, the doors were closed, the lights were extinguished, and the beds were filled. Around the clock, around the years, through the generations.

It went on like that across the entire midsection of the United States, which became not just one region among many, but more of a national reservoir of values and scenes insulated by the coasts. The area came to be called the nation's Heartland. The pioneers first penetrating this vast continent could not emerge from the foggy, gray vapors of the Appalachian hollows to stand by the massive muddy motion of a river to be called the Ohio and gaze out over a land that stretched one-thousand-four-hundred miles before them to the towering mountains of the West, beneath a sky that went even farther, in an untamed climate that spawned humility, with a wind that spoke of powers far, far beyond the control or belief of any puny human. They couldn't stand there before all that, those little folk in their sweaty shirts and stained hats, and think about a tiny tomato patch. Big land, big fields, big thoughts, big deals.

Europe extends to the Alleghenies, Emerson wrote, where America really begins. The Midwest, prairie playground of the glaciers that ground down the old ocean bottom and spread the fertile silt for another use another time, was the real land of opportunity. Untamed, unbroken, and undeveloped, the former shallow sea had a coordinated bounty no mortal could plan. It had the vast stretches of rich soil, in some places seventy-five feet thick. It had the staunch grasses to hold the soil in place, waiting. It had the weather, the long warm days for growing and the long cold winters for resting. It had deep, wet snows which, with the ample rains of summer, would provide the vital moisture for plants. The Midwest had the underground pressure of the ocean and swamps to make the black liquid, black rocks, and black ore for burning and melting someday. It had the inland network of watery flows and freshwater lakes to move this bounty within and without the region. It had the strong sun and the heavy humidity that wraps all living things in a hot, wet towel, and helps rot the refuse from last year's crops. It had the little creatures that fed the big creatures that fed the settlers, and the winged creatures that cleaned up afterward. It had the storms that raged from the west and north, bringing death in winter to the sick, weak creatures and, come summer, the brilliant, noisy flashes from above that ignited the fires that burned the grasses that destroyed the trees that fertilized the soil that still waited. And it had the winds that washed back and forth, invisibly moving the fires and the weather systems from above and warning the creatures below of change coming behind the horizon.

To this stage came the fit. Those who couldn't, didn't. Life there bred a boldness, or bluntness. It was a physical, commonsense challenge, not an intellectual one, a life where drawing rooms gave way to barns. Things get done by doing, plain and simple, with as much applied force as necessary. The Midwest, home to nearly half the gross national product, was a land built for muscular encounters in commerce, in industry, in politics, and in athletics. Hard land, hard fields, hard thoughts, hard deals.

The culture was one of broad pragmatism, a quid pro quo, punch-in-the-nose practicality that showed in the simple talk, the hard work, the utilitarian cities, the plain, fundamental religions, the homespun clothes and politicians. Not subtle, not spectacular, but effective and pretty in its plainness, like the prairies. Wrestling a plow and a team of horses or working with vats of molten steel or rubber or an array of naked auto bodies on moving hooks built more than muscles; it forged a way of thinking that did not prize subtlety.

The Cleveland Browns, the Green Bay Packers, and the Monsters of the Midway, the Chicago Bears. The Chicago Democratic machine. The Board of Trade. Paul Bunyan. The Mississippi River. Big, all right. Even the first controlled nuclear reaction.

The region also bred caution; this morning's sunshine could lead to this afternoon's thunder, or worse. Each year's school schedule contained a batch of "snow days" for the blizzards that closed the roads for a while. Sell some of the corn, okay, but hold some back, just in case, right? You never know. Keep some money in the cookie jar for a rainy day; sure, this year's crop is okay, but next spring could be different.

The diverse Midwest bred leaders who sought consensus. When the surrounding elements are harsh and unforgiving, those within had best learn to get along with each other. And so today, from this one region, comes half the Supreme Court. In the decades since the Civil War it has produced half of the nation's twenty-two presidents and two of the last three. The region has bred a kind of stolid commonsense inventiveness. Illinois's John Deere wondered why a steel plow couldn't be made to properly penetrate the thick soils. Henry Ford wondered why automobiles and automobile parts couldn't be standardized for quicker assembly on a line. Cleveland's John D. Rockefeller wondered what could be done with this oily black goo that oozed up from the ground. Ohio's Orville and Wilbur Wright wondered what would happen if they designed an airplane wing that had more air pressure lifting from the bottom than pushing down on the top. Chicago's William LeBarron Jenney wondered if a building couldn't be designed to be structurally strong enough to stand so tall on a small plot of expensive downtown earth that it seemed to scrape the sky.

The Midwest has also been a social crucible for the country, drawing in immigrants, changing them, and sending the people and products and thinking back out along the nation's spokes to produce long-lasting effects. On the land, of course, midwestern farmers devised strains of grains and methods of producing them that were so efficient and productive, they now feed several lands, and through their export earnings quietly finance much of the country's manufactured imports. The region's small-town values leavened a gawky new land's aggressiveness, producing fears such as those Calvin Trillin, the writer from the Midwest, described with his hypothetical ex-midwesterner in the Big City, "a stockbroker on Wall Street who firmly believes that on the day he starts wearing Italian shoes and drinking foreign wine an old high school buddy from Grand Rapids will arrive to make fun of him for doing so." In its massive urban areas, especially Chicago, a city named for an agricultural product (an Indian onion), the region also produced social activists and decades of social action on both sides of the Pullman riots and the Depression's "penny sales."

The East Coast with New York City was long where the masses passed through, originator of smugness and fading self-importance, seeming seat of power, towers of steel and glass looking out on the ocean and seeing themselves magnified in its reflected natural scale. Here were purveyors of popular thought where the cement climate is dominated by people, many of them midwesterners lost in the crowded numbers who forget their manners amid the rushing decay.

The West Coast was California, land of the Breakfast Burrito, where swimming pools are something to sit around, where grown men wear sunglasses on their forehead and long-sleeve sweaters in the summertime so they can push the sleeves up against the heat. It is a fast-paced place where everyone is famous and nothing is worthless. Life amid the free-flowing freeways and mores in that capital of hot tubs seems to a distant plainer plains to be a ceaseless pursuit of aimlessness where loyalties are changed like jewelry, and everything that doesn't move for a few minutes is named for a dead Spanish saint.

The East Coast's memory seemed to the Midwest always to represent what was; California's Tomorrowland Fantasy World seemed to represent what would be, unless The Great Earthquake came in time. It seemed that the vast Midwest, everything in between, unintentional sifter of signs, the great national balancing wheel, was never much concerned with fads. Fad viruses swept and consumed both coasts on fashionable tides, washing places and people who knew better, standing on little of lasting worth. The plain old Midwest was mocked elsewhere, often by midwesterners who had left, although they seemed to resent the region less for what it was and more for what they were when they were there. But the faraway fads couldn't become trends in the United States until they were accepted by the Heartland, the broad place in the middle with the big hands and the innocent smile, where old values like hard work, patriotism, and neighborliness hung on long after they had been discarded as unfashionable most everywhere else.

In the Midwest there is always the threat that what seems so normal, so calm and peaceful, vital and beautiful—a little cloud, for instance, or a refreshing breeze or a light rain shower—can so suddenly turn ugly, violent, and lethal. Within the cloud, perhaps unseen all along until just before impact, are ice balls that can in an instant smash windows, kill livestock, and turn to pulp the crops lavished with so much attention and money. The breeze can suddenly build speed, knocking down trees, ripping off doors, or, worse yet, flattening the frail green stalks that stood six inches apart as far as the eye could see. The little shower can expand into a full-blown thunderstorm

fifteen miles tall, a moving mountain of black air un-
leashing the same energy as a twenty-kiloton nuclear
bomb, hurling white-hot sparks three miles long at
objects that dare to stand above the prairie soil.

Nature isn't the only thing that hides its feelings in
the Midwest. Newcomers or casual travelers just passing through a cafe conversation could very easily think
the region has no politics, no prejudices, no outspoken
opinions, no alcoholism, no drugs, no nagging wives,
no brutish husbands, no lying children, no racism, no
sexual inadequacies, no financial instabilities, no anger, no suicides, nothing at all wrong really, except the
weather and crop prices.

Public displays of emotion run against the ethnic
grain of independence and self-support that stiffened
the backbone of so many midwestern immigrant groups.
There's some good yelling at home, to be sure, but
more often than not it deals with a random symptom,
not the problem. Midwesterners, mainly the rural men,
keep their troubles and emotions inside like a brewing
rainstorm. The surface looks so calm, until one day,
seemingly without immediate cause, the pent-up forces
become too strong or the will to control them too
weak. There is an explosion, a violent storm of emotion, of wind, of tears, of rain, of violence. It is lethal
lightning that usually nails not someone or something
who "deserves" it but someone or something who
happens to be near enough, warm enough, tall enough,
slow enough, or simply unlucky enough to be around
when the gods above, for their own celestial entertainment, randomly send down death.

It's not that the Little Leaguer deserved death or
did anything to prompt such a final punishment but
just that he or she happened to be on the pitcher's
mound when the cloud's unseen electrical charge grew
so large it struck out for the tallest and warmest object, a blonde head. It's not that the father in the
tractor seat had done anything evil but just that he
happened to be in the driver's seat when the engine's
torque grew too great for its own rear wheels and
flipped its lighter front over to crush the human. Not
that anyone was at fault as the woman stood next to

the pulsing engine when the old casing gave with a powerful crack that drove the chunk of forged iron deep into her chest.

When a string of massive thunderstorms rumbles across the land and a single whirling cloud dips down from the taller blackness as if pointing, it is not the winds themselves that are so terrifying, although they are awesome in their blunt strength.

What is terrifying about these unseen forces is their awful, unpredictable randomness. When the forces are so large and enveloping, there is nowhere to hide. It is a fatalism familiar to North American Indians but foreign to white North Americans. The storm is coming. And if it's coming for you, there is nothing to stop it. It can hit a barn with that distinctive roaring *whumpf,* in a flash turning the modern structure into a splintered pile of wood and twisted steel beams and oozing livestock. It can wipe out one farmer's cropland as if it had been vacuumed, which it had, and leave another's nearby clean and calm. It can lift off the roof of the house, hanging the curtains up over the walls' edges, and leave a baby sleeping peacefully beneath the night sky. Or it can move the entire house four feet and not chip a single cup, although pieces of straw are nailed into the outside walls. It is the uncertainty that frightens. That is why in many midwestern places the hourly newscast begins with the weather. Anything else ranks second. As far away as northern Texas, volunteers still go out from town to watch for the assault of those fateful clouds and to radio warnings to friends behind the ramparts back in town.

Such sudden violence doesn't seem to fit the peaceful countryside, where life and growth come so slowly.

Afterward, of course, the warnings could be seen. It was that time of year. Conditions were right. It had been raining. The government weather service had predicted more. But it was such a normal rain, just a mist really in some places. And then suddenly, without warning, unless you had been looking for the clues very closely, the noisy violence, dispatched from a hidden source somewhere, struck and was gone, leaving behind the rubble and shells of homes and hopes and humans.

There is a rhythm and a balance to the storms, to their death-dealing destruction, to their life-giving moisture, and to the humility they spawn in the teaching words of parents cuddling a frightened child during midnight thunder. If the storms fail to come, according to an even broader rhythm of years, then life becomes unbalanced, the crops fail, the nation's balance of payments suffers, and food prices rise. Midwestern life begins in the overcast spring. It has boundless optimism in the summer sun. And life ends in the foggy fall, sometimes suddenly for the hunted.

Yet these large unchanging rhythms mask gradual changes. The Midwest may have a reputation for resistance to change, but few parts of the United States have shown more dramatic changes over time and none has been able to adapt any better than the agricultural middle, where every state save a handful has seen industry come to dominate or rival agriculture in its economy. Once there was no world market, no grain embargoes, no OPEC, no inflation, no drugs, no floating interest rate, no rising expectations, no liberation movements, no cynicism about government, no need for much money or college, no nothing that a good strong back, a long day's work, a few sturdy sons, an understanding wife, a fear of God, and a few good neighbors couldn't handle.

Once rural and small-town America was insulated from a meaner outside world by distance—self-contained, self-satisfied, self-supportive. Once midwesterners sat on the front porch of a summer's evening, listening to the crickets and each other, visiting with family over coffee or homemade lemonade and neighboring with neighbors over pie so fresh that the crust was still warm. Now each family is down in its air-conditioned, basement family room in their clothes from Taiwan, watching the same television sitcoms as cityfolk, seeing the same violence and jiggly sex, hearing the same news and views.

But across the region dramatic changes were afoot. Self-reliance, the sturdy old virtue, mattered less—in fact, it could be a real handicap in an increasingly interdependent world. Self-contained, self-satisfied, and self-supportive industries such as steel, cars, rubber, banking, chemicals, manufacturing, and farming were being crushed by forces far beyond their comprehension. The ore and coal mines of Minnesota, Kentucky, and Illinois were troubled because the area's aging steel mills were being outstripped by advancing young competitors. The steel mills were troubled because their buyers, the auto makers who created one out of every six manufacturing jobs in the country, were being challenged, also by energetic, imaginative corporate competitors, also from abroad, who knew the American customer better than the local firm did. The American farmers' exports, a national economic mainstay, were being hurt by the strong U.S. dollar and cheaper competition from developing countries forced to export their agricultural products to finance their huge international debts. New financial and physical realities were overwhelming the established ways.

The prairies were built for nomads. First, the Indians followed the buffalo. Later, midwesterners traveled back and forth within the region building the nation's industrial base, always seeking a better job on familiar terrain. And there always was a better job somewhere nearby. During World War II and after, the region's booming factories from Youngstown to Wichita drew millions of workers who got rising wages and overtime guaranteed, and, of course, a better life for the next generation, guaranteed. Many of these new workers were drawn from the farm by the security of a weekly wage impervious to the ravages of weather and shifting government policies. In 1950, twenty-three million Americans lived on farms. Thirty years later nearly four out of five had left the land.

They moved into the diversifying large and small cities, the Minneapolises and Mankatos, the Chicagos and Rockfords, the Akrons and Omahas, discarding the smaller towns like Ruthton as American society has disposed of everything from mining camps to Pepsi-Cola cans. The rural reward for producing much of the country's self-reliance and social values, and spawning and implementing a revolution that tripled production and efficiency in a competitive new world, was that thousands of midwesterners became economically redundant, no longer able to afford to work hard on a farm. It wasn't denying people the right to work. It was something new and worse in a self-reliant land; it was denying a way of life for many.

While the isolated farm certainly remained, its ties to distant places became crucial, increasing the influence and power of outside factors on local life. The price for its increased production was greater specialization and a dependence on others—and their prices. Where once a farmer could grow his own "tractor" (a horse), its "fuel" (oats and hay), and his food, now he needed a powerful green machine from Moline. It needed fuel from Saudi Arabia. And he had no time to fool with a few hogs, a few cattle, and a few chickens for himself; better and cheaper to concentrate on his own speciality—corn and soybeans or wheat or cattle or hogs. In 1974, when soybeans were bringing the farmer nearly $10 a bushel, a new tractor cost $14,360, twelve times its 1950 price. Today the price of soybeans has fallen to around $6 a bushel while the price of the same tractor has nearly quadrupled to $55,000.

While farmers benefited briefly from inflation, which pushed up the value of land and other assets used for collateral, they found themselves in the mid-1980s on the cutting edge of deflation as their interest rates stayed high while their land prices plummeted and their crop prices stagnated. A record number of foreclosures, forced farm sales, and voluntary liquidations spread as local bankers, frightened for their institution's own future, squeezed the local farmers, who didn't see how they could work any harder, get any more efficient, be any smarter. For every seven farmers who went under, one local business folded too. Opportunity was shrinking in the land of opportunity, fraying established values and relationships. It built severe tensions within communities as bank vice-presidents foreclosed on high school classmates, who were doing everything their fathers said would guarantee success.

There was less trust in communities once held to-gether by nods and handshakes. Now storeowners, longtime friends, asked for payment up front, or a letter from the bank guaranteeing payment come harvest. The bank, which was struggling through its own confusing and threatening Darwinian world of deregulation, was leery of many new commitments. In many cases it had been purchased by a larger regional bank that sent in a cost-cutting team from the city to straighten things out, oblivious to the local social bonds. They sent out computer-printed warnings to folks who'd never missed a payment, and were proud of it, just to be sure they wouldn't think of such a thing now. They printed out letters to the grain elevator and the livestock sales barn telling their owners to put the bank's name, too, on any checks for the farmer. The unprinted message in all this was unmistakable: suspicion and fear were being unleashed.

Some families, known locally to be in financial distress, became too embarrassed to attend church. Others dissolved into sleeplessness, bickering, or worse yet, physical abuse of women, children, and even animals. Bankers stopped dropping by the bar for an after-work drink with the boys; the boys would move away from the table as they sat down, and after a few more beers someone might get to talking loudly about hogs and bankers and worms, all in the same breath. Some studies predicted three or four out of every ten farmers would be gone in a few years, which prompted some to quit and many to work even harder. Many small towns became largely collections of the elderly. Mental health centers were busy. So were liquor stores. And the rural suicide rate far exceeded its urban counterpart.

Federal agencies, established as agricultural lenders of last resort, were closing down the same farm operations that had the same land, the same equipment, and the same assets as when the same loan officer first approved the borrowing. The farmers had gotten so good at their business that in some years the government paid them increasing incentives not to farm; the more they didn't work, the more they would receive.

Others, forced to increase their farm's volume to satisfy creditors' demands, found themselves plowing up land they knew was too steep to work, that was subject to wind and water erosion. It was damage, like deficits, that would not show up as serious for some time. Every day another 5.5 square miles of land were lost to urban development, 2,031 square miles every year, nearly two Rhode Islands' worth of space. This new production added to the surpluses, further depressing crop prices.

With so many people leaving the countryside and fewer people working more of it, there was a loss of physical and psychological ties to the land and its values. Some rural school classes went on special field trips—to a working farm—because even though they lived in the country, few of the pupils' parents were actually farmers. Poverty has always dwelt in the countryside, but now there was more, with many farmers needing a full-time wage job to support the full-time farm job. More wives went into town to new jobs. The Midwest to many had become a region with a wonderful future behind it. Farm fathers found themselves giving their boys a different sort of talking to; don't become a farmer, son, it don't pay to work this hard. Some sons listened, some didn't. Giving up and losing out on the land may not have meant much to the world. But it meant the world to them.

Hardship, like midwestern storms, strikes unevenly. But the widespread hollowing out of mid-America, physically and emotionally, left behind decaying towns and neighborhoods packed with thousands of poor who knew firsthand how empty the promise of prosperity was. It left behind Merle Haggard's song asking, "Are the Good Times Really Over for Good?" It left a new name for an old area—the Rustbelt.

"A few years ago," said Duane DeBettignies, the editor, publisher, chief reporter, and owner of the Ruthton newspaper, "in the course of four months we lost Red Lauritsen's lumberyard, the Red Owl grocery, and the Allis-Chalmers dealer, Hildegard Implements. Three businesses gone just like that. And they was all liquidated 'cause there were no buyers. We used to be up and thriving. But now I'm thinking about the paper. I'm fifty-four, I don't know if I wanna be scrambling around like this in a small town like this for another ten years. I just don't know." Time was, in fact, when Ruthton had three farm implement dealers, a drugstore, a movie, two car dealers, a furniture store, three restaurants, a roller-skating rink, a creamery, a doctor, and a dentist.

Now Ruthton's Doc White, the dentist, has moved away to the Twin Cities. Doc Sether, who had passed out the same corny jokes and tasty lollipops to generations of children, well, he died years ago. No one ever took over his clinic, so one day it just closed, and people had to go elsewhere because the younger doctors don't make house calls. The new doctors in the modern low-rise building with the tinted glass and the swishing air circulation system seem younger and better-informed about medicine, or at least they have an answer for everything. Their magazines are just as old as the old doc's. But nurses catch the eyes of these young fellows glancing quickly at the name line of their next patient's file as they enter the examination room with the same cheery greeting as when they entered the other room across the hall four minutes before.

Now there are fewer customers for everybody. People have cars and trucks that take them to a shopping mall that offers larger discounts on larger inventories and customers buy from a salesclerk they never saw before and will never see again.

Not being the county seat, Ruthton, out on the northern edge of Pipestone County, has seen its population nearly cut in half over the years, dwindling closer to 300 now. One in ten of the county's residents moved away in the last decade, leaving behind only 11,690 souls in an area of about four hundred square miles. Tractors no longer park along Ruthton's main street.

The townsfolk still wave at a familiar pickup truck easing down the street. And some folks still stop by the little post office pretty near every morning to pass the time of day. According to midwestern small-town tradition, all these conversations must be noncommittal, nothing controversial, nothing outspoken or too forthright. Folks has got to come around to new thoughts or ideas at their own natural pace; no amount of yelling will make the corn grow any faster and the same goes for new ideas taking root. Just common courtesy, you know. And a desire, sometimes even a desperation, not to rock the boat. As most everyone knows in a small town like Ruthton, a small boat rocks much easier than a large one.

There are no longer school holidays in early May so the boys can help in the fields with the planting and the girls can help in the kitchen. The school bus drivers are women now, some with little preschoolers in tow to save on the baby-sitting; their farm family needs the extra income, and it's one formerly male job the women can do that the men don't seem to mind too much about. The female drivers yell a lot at the noisy kids bouncing in the back along the rural roads from long driveway to long driveway. But the women's yelling just doesn't have the same powerful effect as some years back, when their beefy male predecessors would just pull the yellow bus over to the roadside, pick up a wooden paddle hanging by the front door, and wordlessly stride down the suddenly empty bus aisle to whip up a little pink silence back there for a few miles. That wasn't considered physical abuse in those days; it was a way to teach "discipline." It was called "respect for your elders" too. But it wasn't really. It was respect for powers far larger and far more powerful than yourself. Those men, fathers all, had versatile hands and understanding eyes, and like everyone else in the countryside to this day, they wore heavy boots. They didn't choose to talk much. They didn't pay much attention to disciplining the girls. Sometimes the little boys, when they stopped bouncing around for a minute, would see a flicker of rage behind those men's eyes. An angry storm, controlled for now. But if someone had started talking about "creative expression" and "positive reinforcement" in those days, the men's eyes would have turned up toward the ceiling as if searching for cobwebs.

There is today still plenty of simple philosophy about. "We get in trouble with our attitudes and greed," said Mr. DeBettignies one day as he worked on his recalcitrant printing press. "I mean attitudes about each other, different people, and countries. People has gotta stop once in a while and be thankful for what they've got—health, family, a roof, food—instead of always thinking about what they ain't got. You know, you don't have to be wealthy in dollars to be wealthy in life." But there are new currents even in his life. His wife, Marlys, has to work now, or at least she chooses to; she's a registered health service nurse for the government, one of a few sectors actually increasing employment. Of their three children, none settled in Ruthton.

"I was born and raised in a small town," says DeBettignies. "I loved the freedom from the drag of life, all that city traffic and hassle and run, run, run. Ruthton was a good place to raise children. I got no criticism of kids when they leave. Small town or large town, you gotta go where you think you can make it, even if it is a desert." The family still tries to get together once a month or so. It's more than a lot of young people see their parents, and grandparents see their grandchildren, although it's a lot less than everyone is used to.

When Sharon Fadness's parents split up years ago, it was not all that common around Ruthton. D-I-V-O-R-C-E was still a word parents spelled out in the presence of little children, and it just wouldn't do for both parties to remain in the same area. So her father left town for good. Every summer from the age of six on, Sharon would be sent off to visit her father and his parents in Missouri. There, and later when she attended college in California, Sharon was routinely exposed to different people and their different ways of looking at things. It wasn't the difference between, say, a Madison Avenue executive and a Maori tribesman. But she came to see for herself that Missouri's way, California's way, or even Ruthton's wasn't right or wrong, just different. That can be a threatening perspective in a place like Ruthton, unless you guard your words very carefully to disguise a broader experience.

Still, an aware family can take socially acceptable steps to insure that an older son is exposed to life outside surrounding counties so that if he, like his mother, does return one day and does take up livestock hauling like his father, it is because he chose to, not because that's all he knew. So Alan Fadness, the maturing teen-ager, a straight-C student who got excited around his father's shop and trucks but left his motivation outside the local high school door, was thinking about a stint in the navy, a ticket out for a while. "If all you see around you is all you think there is," said his mother, "well, kids need to go out and see the world a bit away from home to make sure what they want." Not many young folks were sticking around Ruthton anyway. The opportunities there were dwindling. Hope was drying up. And then came the rumor that the bank was being sold.

The speed of a particular rumor in a small community is directly proportional to its surprise and perceived import. The rumor flashed around very quickly over coffee cups and phone lines that after all these years of quietly running the only bank in town, it seemed that Clyde Pedersen, too, was pulling out. He was selling the Buffalo Ridge State Bank. Clyde was a known commodity to everyone, not exciting but not threatening, comfortable. He hadn't created the bank; he'd inherited it. Now he was selling out to some easterner, that big tall fellow who was seen striding around town a few times. How he'd ever heard of a place like Ruthton beat everyone. It's one thing if Alan Fadness comes back to town; he belongs. It's quite another if someone else from somewhere else actually chooses to come to town, especially someone who uses a middle initial. The new owner was married, they said, to some woman from Phildadelphia. Had a kid, a boy, they said, probably spoiled rotten like a lot of those city kids. They must be rich, awful rich, to buy a bank. In some places, even in Ruthton in another era, such fresh blood with foreign ideas might not have been threatening, might not have spawned incoherent fears of new financial priorities, different standards, stricter accountings, new ways. It might even have raised fragile hopes. As usual, DeBettignies heard the rumor quickly. His first thought: "What will happen to my loan at the bank?" His second: "Why would anyone ever buy in here?"

3

THE BEGINNING

Rudolph Hamma Blythe Jr.'s bubbling enthusiasm filled the car as he drove along the open prairie highway with his family. To those who didn't know him well, Rudy's size magnified everything he said and did. Even his friends would wince sometimes at his powerful physical presence. They had since his days on the offensive line of the high school football team when his schoolmates back in Philadelphia voted him the loudest member of the class of '59. When you are six feet four inches tall, when you have a full tenor's voice, and when you carry 250 pounds, sometimes more, a whisper from your mouth can be heard as a shout.

Rudy was full of optimism and awe and as expansive and excited as a little boy going to see his first professional baseball game. Rudy was taking his family to Ruthton, Minnesota, to see the small-town bank he wanted to buy. It had great promise; he just knew it. The town had a lot of potential too; he could tell just from walking around. He'd heard that the bank's present owner, the son of the previous owner, hadn't been all that interested in the community, so he wouldn't be a hard act to follow. And the price seemed about right. It was a lot of money, but Rudy had managed big sums in both his bank investment and trust jobs in Minneapolis and Des Moines, and the interest rates would be going down soon, no doubt. He would get a large loan now with a floating rate and cut his costs with a new loan at a lower percentage when the rate fell in a few months or so. Rudy was very eager to realize these plans because, nearing the age of forty, he had just about had it working for other people in the army and business, where bozos rose according to their incompetence.

After eight years of marriage Susan Blythe could tell when her husband was happy and excited. He would talk very rapidly and a little loudly, the words and thoughts pouring out from on high. Although she couldn't always understand the forces behind his happiness, it usually made Susan feel good too, or more comfortable. Life was much easier and more pleasant free of six-foot-four-inch scowls.

"It'll be just great!" Rudy Blythe said as he slowed the car. "Well, there it is!" His wife looked all around her, through the windshield and out the side windows.

"There is what?" she said.

"There's Ruthton," her husband replied, pointing ahead. She looked in that direction. There across the flatness of some muddy brown fields stood a large clump of bare trees with some discolored houses in scattered disarray and a rusting water tower standing overhead. Oh, God, she thought.

Rudy Blythe was not a subtle, graceful person in those days. Big, tall people don't need to be. And he never was. Each morning back in high school in St. Davids, Pennsylvania, along Philadelphia's famous Main Line, he'd come running over the hill toward the quiet male crowd of serious, suit-clad commuters waiting decorously on the train platform with their morning newspapers carefully folded to the business or sports pages. Young Rudy was often running late in those comfortable days of the mid-1950s, when a grandfather named Ike presided over the prosperous land.

Rudy went to a prestigious private high school, Haverford, along with several hundred other sure-to-be-successful scions who had to wear shirts and ties and jackets so they would become young gentlemen. He commuted to school by train and usually he would come running over the hill just as the brilliant light of the engine neared the affluent community. A big youngster—he weighed nine pounds four ounces at birth—Rudy would be running and carrying his sports coat and a stack of books. And a book or two or perhaps the whole stack would slip from his arms and fall on the ground, spilling papers and geometry tools about. Rudy would stoop to pick it all up in a hurry. The wrinkled tails of his white shirt would slip out as he bent over, and a number of the neat businessmen, who really weren't reading the paper anyway, would turn and watch and smile silently and perhaps exchange knowing looks with one or two other businessmen standing nearby—and perhaps unobtrusively reach around and check their own shirttails too. These men didn't know each other's names, never would either. They just knew the same familiar faces at the same familiar places each morning en route to work—visual friends for ten minutes a day, five days a week, fifty weeks a year. But these men had been young too once, probably. And they had been awkward. And seeing that unorganized young fellow spill his books and sharing that moment with someone else made them feel good. Life was still going on as usual, of course—why wouldn't it? And it made them feel good too that, by comparison, they were no longer so awkward and so young and, by appearances anyway, so unorganized. The hissing train would arrive then and they'd pick up their leather briefcases and step properly toward the dirty tracks and the oily wooden ties and the sliding door, which was closing as the disheveled young schoolboy leaped inside just in time.

Rudy was an average student—lots of C's, some B's, and one or two A's over the years, enough to keep him hovering around the middle rankings of his class of 150 in a school where going on to college was a given. Without much hair—the crewcut was much in fashion then—young Blythe's appearance was forceful, with his broad nose, his big ears and thick lips, his intense eyes, and those big shoulders, seeming even more immense under his football pads.

All the guys had a nickname or two, anything in those days to avoid using the Jr. or the Roman numerals parents tacked on their names. James Ray Shoch III, who played next to Rudy on the line, was Jamie. Bernhardt Camille Smith was Smatls. Edwin Corning

Sinkler was General. Lance Jonathan Boerner was
Wedge. Young Blythe was Rudes, Root-toote-toote,
or, more respectfully, Animal for his build and strength
and ways. So at that sunny Saturday home game when
Rudy determined that the spectators were not cheer-
ing loudly enough for his teammates on the field, he
turned around. The big youth stood up on the bench.
The big voice ordered everyone to cheer more. And
the little crowd did what it was told.

Rudy loved his world of males and male groups. He
always would. The sense of acceptance, of belonging,
of being one of the guys. Sweating together, swearing
together. Guys can be cruel, all right, establishing
their pecking order with verbal jibes and physical shoves
to see who will be the leaders, who will be the leaders'
assistants, and who will be led. The roles weren't
assigned. Guys just fell in where they felt most comfortable and where peer pressure permitted. Rudy was
in any group he wanted to be in. He could be a leader,
but he was usually a leader's assistant. He was most
comfortable if the group revolved around something
physical like sports—football, wrestling, swimming, pok-
ing around with his pals. Males have a way of automatically awarding some deference to their larger peers.
It need not always be a group based on physical en-
deavor. Rudy was in the Auto Club and the Glee
Club, too, where he would stand out in the ranks of
serious singers due to his developing penchant for
plaid clothes. If the group had to do solely with ideas,
Rudy was not all that attracted.

He was not a loner, nor was he the life of the party,
although he could be the center of it. Like the time a
large Rudy swept his dance partner around the floor,
the only one to do a polka, and the others got out of
the way to see better—and to protect their own safety.
He was to seek out such groups throughout his life.
Rudy had to be around people. In his groups everybody knew Rudy Blythe. He liked everybody, and
everybody seemed to like him. They better, they
thought.

In his family, too, male companionship was important. Many years later on a quiet evening with friends

he might recall in passing how much he had looked forward to Saturday mornings in his childhood. Not because of the radio adventure shows or, later, the westerns on television. On Saturday mornings his father, a leading research pharmacist who revolutionized the drug industry by inventing the time-release capsule, would get up with Rudy. He would make breakfast for his son and they would play games. That was the special time for them. Teen-agers often shun association with anyone as gross as a parent. But even as a youth Rudy would talk proudly to peers of his father's accomplishments. They would pick up on his serious tone and no one would make the usual mocking quip. Over the years, young and not-so-young, at home and abroad, Rudy would seek out male associations with the plodding, powerful determination of a defensive tackle. Where the groups existed, Rudy became a member. Where such groups did not exist, he helped create them. It was very important to him to belong.

Those high school years were also spiced with the usual teen-age tomfoolery. There was the time when he and a few teen-age friends paid a visit late one evening to an empty estate for reasons that seemed challenging at the time. There Rudy climbed a tall metal statue, and his big, powerful arms accidentally removed the head. That happened just as a policeman arrived with a few questions about the statue, the head, and their names, and a curiosity about the contents of Rudy's car trunk. Except that Rudy had lost his trunk key. When it was finally opened, the trunk yielded not beer, as the officer had suspected, but just the old tennis shoe Rudy had been looking for. Or the time Rudy visited an aunt near New York City with a classmate. They went to Greenwich Village, two young men barely old enough to drive, and walked into the first dark place they figured their parents would not want them to go: a beatnik coffee house. There they drank a few cups of that stuff, bitter and dark, smoked cigarettes, listened to some poetry, and discovered that boredom could easily flourish outside suburban Philadelphia classrooms.

Just before graduation Rudy and his friend Tom

Widing determined that Rudy needed a new car to replace his aging eight-year-old Mercury. The new vehicle would be more fitting to their image as young men. Perhaps a spiffy Corvette would do. Over a beer or two it was decided that Rudy would approach his father that very evening and broach the subject by saying he wanted to have a heart-to-heart talk about cars. It sounded good and serious, and a rare sense of teen-age confidentiality at home always gets the attention of dads. The two young men went to Rudy's house on Hilaire Road. Tom waited in the hall. Rudy returned sooner than expected. "I blew it," he said. "I walked in and said, 'Dad, I need to have a car-to-car talk about hearts.' "

Rudy was not very big on handing out philosophical suggestions, even the kind of sophomoric statements that clog generations of senior yearbooks. He tended more to the simple, the practical, the straightforward. Although he was to become fond of books in later years, English and correct spelling were never his strong points. But a lot of the guys would want to write in his yearbook, creating a link with the big man. "What does the future hold?" pondered "General" Sinkler before answering himself, "Who knows!" William Percy Arnold III told him to work hard and "you will go very far in life." John Strong Bevan, better known as Bevo, recalled Rudy's dogged determination as an example for underclassmen. William H. Ewing thanked him for joining the Columbia Record Club and earning Bill a free record. Wedge Boerner told him not to go too hard on the farmers at Franklin and Marshall College out in the sticks of central Pennsylvania, where Rudy would begin college the following school year. Charles Crooke Auchincloss II—Awk to his friends— reminded Rudy to control his temper.

Pete Ward remembered the bang Rudy got from life in those days and kidded him about being overenthusiastic, especially joking during the fourth study period on Friday mornings. Fridays are a tempting time for students who look forward to the weekend so much. Rudy liked weekends, not so much to catch up on unfinished work from the week as to forget completely for two days the responsibilities of the other five. Hard work wasn't always necessary; good times were. One college friend remembers going to New York with Rudy once for a debutante ball. A girl's parents had hired a limousine to collect some of the dates. As the shiny car slowly weaved its way through Manhattan traffic, Rudy stretched out his long legs—and still couldn't reach the plush back of the front seat. "Now," he said, folding his hands behind his head, "isn't this the good life?"

The summers of the good life were welcome respites from the busy eleven-hour days at high school with all the required athletics and activities. The Blythes were solidly middle class. Coming out of a small New York town, the son of a farmer and a graduate of Columbia University in the midst of the Depression, Rudolph Blythe the father, or Big Rudy, had started the research operations at Smith Kline & French, which he turned into a large success. Although Mr. Blythe only got one dollar from the company for the patent on his time-release capsule—first applied to amphetamines before leading to the invention of Contac—he was paid quite well and was very prominent in the industry. There was a little family money, too, and over time Mr. Blythe was able to parlay his funds into a profitable stock portfolio, an activity that dominated many family discussions and became a satisfying pastime well after his retirement. Vacations, like most everything in the Blythe family, were usually put to good use, combining relaxation with tax-deductible travel. Sometimes they'd visit relatives in rural New York State. Rudy especially liked those small towns with the surrounding livestock and all the pets, where the routine seemed more relaxed than life around Philadelphia. During those visits he first began talking about country life.

Rudy's mother, who was crippled by a childhood bout with polio and channeled much energy into painting, believed strongly that it was good for young men, her young men, to work. To put their lives in order too. Hers was not a request. Hard work, no loafing. It was the American way, the right way. It always led to success. Look at her husband. Her relatives. Herself. At fourteen Rudy, her younger son, was a busboy in a Philadelphia diner, which enabled him, a lifelong dessertaholic, to bring home leftover pie some nights. True to his developing form of scrupulous honesty, Rudy would then file his first tax return in 1956, reporting $4.30 in taxes withheld for the government on $129 in tips, some of which was also withheld by management to pay for broken dishes. Later, Rudy would demonstrate toys. He would sell Fuller brushes. He would usher at the movies, where he got in a little trouble with the manager once for creating a problem with Gina Lollobrigida. He misspelled her name on the marquee.

Sometimes Rudy worked as a waiter or a barman at one of the private parties that lit up the large homes along the Main Line. Half the time it seemed Rudy was working at these parties and half the time he was an invited guest. He knew the hosts and guests through his own well-connected parents, although his mother had warned him about acting like a guest when he was really an employee. One night, at the debut of a judge's daughter, the judge himself asked Rudy, the waiter, to bring him a scotch and soda in the den. When the young man obediently complied, he remarked knowledgeably about the sailboats in the paintings hanging along the well-lit, wood-paneled wall. Impressed, the judge asked the respectful young man some questions. The two got to talking, and the judge, quite naturally, asked the young man to sit down, and they talked for the rest of the evening. That was how Rudy Blythe once got paid by the hour for being a guest. He could do that if he wanted, when he turned on the charm. He could overcome that latent, explosive impatience that lurked so close to the surface. He would develop this sense of finesse, of gamesmanship, of getting people to do what he wanted without their knowing it, or seeming to. Sometimes what he wanted them to do was fun. Sometimes the fun came more from the maneuvering. Sometimes being angry got him his way. Sometimes it required being nice. His mother remembers well the day she visited Rudy's first grade class at recess time, only to find her son absent from the playground. Instead, Rudy was indoors with the teacher, his head in her lap while she read him story after story. Little Rudy had complained of feeling ill enough to skip recess but not sick enough to go home.

Rudy Blythe wasn't too happy in college at first. For one thing, he had to study foreign languages, never his strong point. There was Spanish and then Italian. They required diligent application and memorization, and their usefulness seemed so distant. Rudy always said that Franklin and Marshall was a pseudo-Ivy League school, and one thing people remember about Rudy Blythe was his distaste for pseudo-anything. It had to be genuine. His father had some genuine connections at the University of Iowa, two days' drive away. The school had a good reputation, and Iowa City, a lovely, well-organized university town plunked down amid the practical cornfields of eastern Iowa and the culture of Grant Wood, was largely free of pretension. There were always going to be a few who smoked pipes a bit ostentatiously and talked about poetry more than seemed necessary. But generally Iowa City was the kind of place where most men who wore leather patches on the elbows of their tweed sportscoats did so because they thought it actually prolonged the life of the garment.

Rudy transferred there for his junior year. His parents were worried about their son's acceptance in a region so far from home. They asked a friend, an academic dean, to check on Rudy occasionally. The first year the dean invited the young man over for an evening to help decorate the family's Christmas tree. Rudy helped decorate the tree, then promptly excused himself. He had a date and a party to attend. The dean sent word back to Rudy's home that the young man had settled in just fine.

Sometimes en route to a convention out west, Rudy's father would stop in Iowa City for a day or so. Those were fun times, almost like Saturday mornings again. If his mother came along, it became a formal parents' visit, and she knew it. With just his father, Rudy was more comfortable. He'd show him the campus, take him to parties, have friends in to sing around a guitar, and serve pitchers of beer in between his shifts as a waiter. It wasn't really clear what young Blythe was doing with his life in those days. He was obviously happier, but he wasn't dedicated to anything.

His parents received a phone call late one night. It was Rudy—they called him Dolph—announcing that he had decided to drop out of school and enlist in the army. His parents were not exactly ecstatic, as Rudy may well have anticipated, considering the timing of his call. His parents had already gone to bed when he phoned around midnight. They could only make unorganized argument after unorganized argument against his move, passing the phone receiver back and forth between their heads on the pillows. They knew their son had yet to learn how to fully apply himself. He seemed to be drifting, but drifting in college was better than drifting God-knows-where. At least being in college implied movement toward some goal. They were worried that Rudy would not return to college after the army. "Mother," said Rudy with an unaccustomed air of authority, "that's why I'm dropping out now, so I can do better when I do return." After a while both parents, tired, gave up on dissuasion. "Well, it's your life," they said, which it was. Mrs. Blythe formally accepted the idea on one condition: her son must promise to write home every week, which he did.

Many young men have grown up in the military, or wanted to. It provides an imposed discipline, a khaki steel framework to hide in and to achieve on. Its stripes, bars, and stars establish rankings for members to polish and resent. Jammed together in small units, considerable camaraderie emerges. There are rules that can be broken and rules that cannot be broken, and members practice differentiating them. The military life provides fodder for fables about military exploits and conquests of another type. For many, too, the military is a free ticket to new places and experiences, which never appear on the honorable discharge certificate but nonetheless mold the later preferences and style of the former recruit.

Rudy, who as a little boy had posed for World War II family snapshots in a soldier's uniform with a toy rifle, announced to his father that he would be the best damned soldier the army had ever known. He wasn't. But in typical fashion Rudy tried to make the most of a situation. In typical fashion he didn't quite succeed.

"Rudy," one friend recalls, "was always a plunker, never the star."

He originally wanted to join the army's security service. There was an air of eliteness about it. And they liked large men. There was some minor trouble brewing in a place called South Vietnam, and President Kennedy had dispatched a few security advisors to take care of the situation. Electronics was also an emerging possibility, but looking at his school record—and the army being the army—Rudy Blythe was sent instead to foreign language school in California en route to a foreign listening post in Eritrea, Ethiopia. After three unsuccessful months of nonstudy Rudy was transferred into an accounting course. He wouldn't be a listener at the listening post in Eritrea. He would be an accountant.

He then had an opportunity to apply to Officers Candidate School. Rudy already was six months into a thirty-six-month hitch, and the thought of extending his enlistment just to be an officer was not appealing. Anyway, officers, as a group, are often resented, especially by those who believe they could have or should have attained that higher rank themselves. Rudy could take orders, although he was starting to think that maybe his ways might be better. What bothered him the most about officers was the pseudosocial rankings they attached to their higher status, as if they were somehow intrinsically better than the enlisted men. Rudy often recalled in later years an incident when an enlisted man's child was barred from the party of an officer's son because it was being held in the officers' club.

Rudy loved his time in Ethiopia. In many ways he

was his own boss there, running the accounting branch for a base of some thirteen hundred men, native and American. And the primitive surrounding country was just about as far away from St. Davids, Pennsylvania, as anyone could imagine. The rural life held an increasing appeal for Rudy, especially if he could return to his comfortable barracks every night. The country had its own rhythms. It was natural, not phony. Someone coming into the country from the city would have a built-in advantage, it seemed. There was plenty of room in the country, room for a big man and his developing desire for independence. A person who wasn't necessarily a star somewhere else could stand out in the rural arena, could have some noticeable impact and perhaps more respect. People would notice him. And he would notice the notice and feel good. One time at an African rodeo, Rudy, the large white foreigner, and a game one at that, was invited to ride a bull, just like a real cowboy. When tall Rudy with his tenth-of-a-ton weight climbed on the scrawny creature, the bull fell down. Rudy loved that story. It had everything—being in an exotic land, being invited by the group, being a sport, being big. He told the tale often for years.

Rudy regularly wrote home, as promised. But he never promised what he'd write about. His missives were generally boring, full of military routine and an occasional request for something—some film, a tent. They were generally devoid of feeling or passion. He did this. He did that. He went here. He went there. The mechanical notations one includes to fill up a required letter. The only time any enthusiasm crept into his letters and diary was when he was out sightseeing and when he started counting down his last 142 days in the army.

Rudy liked being a tourist. He was by all accounts pretty good at it, too, methodical anyway. He would fly into someplace—he favored African destinations— and find an out-of-the-way hotel, not a dump but not the Ritz. Shunning souvenirs, he would buy a bunch of guidebooks and brochures. He would make note of every expenditure, down to the smallest local coin.

Back in his room he would read the literature closely.
Then he would systematically set out to visit each
place, one after the other, and reread the appropriate
sections from the books. He would stay in a city a
week and absolutely cover it. Then he would ship the
books home for later reference.

Sometimes he would get to feeling sorry for himself,
as lone tourists do at times when they see laughing
groups going off after a tour to have dinner together.
After visiting a doctor for a minor ailment in east
Africa, Rudy once wrote, "I trotted off like a good
boy back to my empty life." During those years he
was drinking a good deal, a steady habit that he firmly
decided to break years later. "I drank enough in the
army for a lifetime," he would tell friends while sip-
ping a soda.

Rudy began to change, to mature. His reading material moved from Ian Fleming's *Goldfinger* to William
Faulkner's *Intruder in the Dust*. In some ways he remained the same old Rudy. An army friend, who
loaned him a room one time, left Rudy a friendly,
pleading note: "Please help me keep this room clean!!!
From knowledge we get sympathy."

But on one of his journeys Rudy encountered a
Catholic priest, Father Hugo, who ran a local orphan-
age. Rudy became consumed by the Father's cause for
a while. He spent much of his free time there, talking
with the priest and playing with the children. He drove
about the countryside with the brown-robed priest in a
battered Volkswagen. He badgered his military pals
for donations. He even wrote his parents about it.
They talked to a few friends, got some old clothes
together, and sent them off to Africa through the U.S.
military postal system to be gratefully accepted by
Father Hugo.

Rudy's photographs—he seemed to be taking pictures of everything—showed the changes too. He always used a camera as an instant notebook, a recorder
of interesting sights that didn't take much time and
didn't require tedious writing and thinking. His color
slides were full of people, a few friends, mostly strangers. They were caught in their everyday routine and

pleasure-seeking in the streets, in the parks, in the markets, and on the benches and beaches. A mountain with a few people dwarfed beneath it. A dirty foot on a moss-covered step. Ethiopian porters with a full load. Some young white women in miniskirts crossing a street. A man in a bowler. There were many children—he was drawn to their faces—in these filmed notations, sitting, looking, crying, feeding pigeons. Women talking, rubbing on suntan oil, or dozing on their backs in the sun, the top of a swimsuit having fallen down a little, perhaps inadvertently. All these fragments, 1/250th of a second of so many lives caught forever on film and safely stored by a stranger on minute, numbered shelves in little metal boxes a world away, unbeknownst to anyone but the big photographer.

After he was discharged from the army in 1965, Rudy remained in Africa. He wanted to see more of that rural, primitive continent. For several months he hiked across parts of it, took trains and planes, hitch-hiked, and rode the local buses—mobile Old McDonald farms jammed with people and animals and boxes of all kinds, all smelly. He went rock-climbing, assaulting the rocks, puffing and pulling his way up the crevices and paths and slanted plains. Tiring and exotic and fun. Then one night after a very long day on the hot mountain trail two porters came to camp. They spoke not a word; Rudy's foreign language now anyway was numbers and pay vouchers. These little guys motioned for him to sit down. Then they began to wash his feet, respectfully removing the dusty boots and the dirty socks, rubbing his tired toes and ankles and soaking them in warm water, like servants. It was a very strange feeling. Strange, but nice. Something biblical. It was a little embarrassing, too. But it also made Rudy feel great, no, superior—and not just in his feet. He liked that sense of power. I must admit, he told himself, as if a little voice had been chiding him inside, my feet do feel better.

In Kenya he had a brief, minor flirtation with a professor's daughter who was passing through. He fell in with some Britons who introduced him to something called rugby. They clamored to have him on their team. Rudy absolutely loved it. No flirtation here, just instant passion. Like football without pads. Rock-'em-sock-'em, knee-in-the-face, dirty, sweaty fun. Sheer heaven for a big tackle from American football. In rugby he actually got to touch the ball sometimes. If you want that goal badly enough—need it—you can get it. And bloody noses become badges. And afterward, as a team member, he was included in the parties, all the talking and listening and drinking and eating. The group. He felt so at home so far away.

After some months Rudy went back home, through Europe to Philadelphia to Iowa to the university again. Earning money as a bartender at Joe's, he plowed through the academic work with renewed energy and a new perspective. He drew on his time and interest in Ethiopia, writing his thesis on the economies of undeveloped nations. Rudy had a new interest, of course—rugby. His enthusiasm was so persistent and so infectious that he founded the University of Iowa Rugby Club, an achievement and contribution to collegiate society so satisfying and so important that Rudy listed it on his résumé for years afterward. Two years later, graduation came on a lovely June day in the steamy field house. There was the document to prove it: Bachelor of Arts–Economics. He had done what he promised.

Rudy returned to Philadelphia. A girl friend had gotten him an interim job at her father's company. Rudy worked on the shipping dock, loading and unloading trucks with the burly crews. He got along with the other workers, who had not attended Haverford School, Franklin and Marshall College, or the University of Iowa. That first day Rudy forgot to bring a lunch, which didn't matter since everyone on the loading dock routinely shared and traded things from their lunch pails. The second day Rudy did bring his lunch. Just like the guys, he passed it around, a big tub of cottage cheese sprinkled with celery salt. And what is this delicacy? his coworkers inquired, only they used other words. Just then, the company owner walked through.

"Hello, Rudy," he said.

"Hi," casually replied the newest employee with the

least seniority and the most eyes suddenly staring at him. He who no longer seemed to belong.

Soon Rudy enrolled in Temple University for a master's degree in economics. He led a busy life in those days and had a full social calendar. Some schoolmates were still in the area, so there was a network of friends. There was a string of young women, too, smaller, dark women attracted to this large, tall, jolly fellow with the lovely voice who liked good times and could use a little feminine guidance on his selection of clothes.

Immediately upon his return, Rudy had also found the local rugby club. He played hard there on the weekends; sometimes his personal life seemed built mainly around the club, its games, and its doings. He told friends he was coming to like postgame team parties as much as the playing itself, sometimes even more.

Rudy was also getting involved in politics, another group to work with, to party with, to belong to. His father had been something of a Democrat early in life. His mother was always a staunch Republican, and always would be. She sometimes joked that her mission in life was to convert her husband to the true way. She felt she had succeeded. She certainly had with both her sons. Both Rudy and George worked in the trenches of political organizing, envelope-stuffing, voter-phoning, door-to-door pamphlet-delivering. Both would run for minor political office. Both would be firm Republicans. Rudy gravitated to the Rockefeller side of the GOP. He always called himself a fiscal conservative and a social liberal, that is until, years later, he stood in the check-sorting room of his bank's basement and saw where those socially liberal government checks were being cashed. It wasn't at the children's clothing store.

Rudy wasn't on many winning Republican campaigns in Philadelphia, but he met interesting people through his work, including one nice young woman from Minnesota. Mann was her name, Somebody Mann. Was it Barbara? She worked for McCarthy in '68. Her father was a judge or something out there. Rudy would never

know either Mann well. But the gray-haired father
would come to know Rudy Blythe intimately, not personally, but intimately from grisly photographs and
from colored tales that people would tell him one day,
under oath.

Rudy went to school at night. By day he did his
research and homework and, when called, fit in a little
substitute teaching to earn some money. One morning
Great Valley High School in Malvern, Pennsylvania,
called him. He was strolling through the teachers'
lounge there during a free period when he noticed a
teacher grading papers. She had removed her heavy
ball earrings and was idly rolling them in her hand as
she waded through a pile of themes.

"You remind me of Captain Queeg," said Rudy,
the ex-army accountant, pleased with his reference to
the tyrannical captain in *The Caine Mutiny*.

"Don't you mean Captain Bligh?" said the English
teacher, confusing her tyrants. A pleasant conversation ensued until the class bell rang. Rudy, never
really very good at names, promptly forgot hers. Days
later, he sought her name and phone number from a
fellow teacher who sought permission first from the
English teacher.

The woman was Susan Slater. At twenty-four she
was three years younger than Rudy, around eight inches
shorter, and, in those days, at least a hundred pounds
lighter. A small-town girl from upstate New York,
Susan had been born in Dunkirk, hard by Lake Erie.
It was not far from her favorite childhood attraction,
Niagara Falls, the big, tall cliffs where all the beautiful
blue water swept so silently, so swiftly over the rocks
at the little child's feet before suddenly thundering
over the edge to crash into the white mist far below. It
looked so innocent and soft from afar.

Susan's father, Wilson or Bill, and mother, Alice,
had carved out an average, comfortable, if at times
itinerant, middle-class life for themselves and their
children, Susan and Charles. A civil engineer educated
in Cleveland, Mr. Slater worked for the Erie Railroad
when trains were a part of every traveler's life. Mr.
Slater and his family moved along the rails, on their

company pass, from modest home to modest home around the East and the Midwest, to whichever small city the company decided needed his talents. It was good, decent, steady work in good, decent, steady places like Dunkirk in New York, Morristown in New Jersey, Niles in Ohio, and Huntington in Indiana, places with fifteen thousand or more good, decent people.

As a youngster Susan was very shy and submissive, but one hot day after a hard play at a Niles playground a very thirsty little Susan walked toward the drinking fountain. Other children were in line and when Susan got near the front, still more crowded in to sip the refreshing liquid. Then others came. And, as her mother watched, Susan stepped back. There seemed to be an inexhaustible supply of other children ready to push their way in. Susan waited patiently, perhaps fearfully, for ten minutes or so. And then the little girl had just about had enough, or rather, she had not had any yet. So with a good deal of noise and some authoritative pushing, she muscled her way to the front and drank her fill while everyone else waited. The Slaters told Susan and others that story often over the years, a family fable illustrating a crucial change, even if maybe it didn't really happen just on that hot day at the playground. Mrs. Slater still believes that was the last time Susan, the child or the adult, was pushed aside.

She had a determination to be heard too, which perhaps grew from having to stand up for herself in a family with two tall men. One was a six-foot-three-inch, 240-pound younger brother who was a Little All-American offensive tackle in college, and the other was a six-foot-one-inch father who unconsciously taught his daughter to have the same kind of firm, outspoken opinions that he expressed out on the rails or around the dinner table before he watched the evening news with Douglas Edwards or John Cameron Swayze. Susan was taught to be a little forward about what she wanted in an era when outspoken females were not the norm. Being straightforward, getting things organized, and setting out to do what seemed necessary had its early rewards, too. Once she decided she wanted a beautiful new red bicycle, told her father that she wanted it, and her father got it for her. A lasting lesson. Speak up. All he can say is no. If you don't offer your opinion, how are people going to know it, assuming they want to know it, of course. And why wouldn't they? Susan was ready to hear theirs, and perhaps to change her thoughts should someone be persuasive or threatening enough. That is the way to work things out, the way the Slaters worked things out. Everybody had to know that.

Susan was popular in school, an outgoing, easy talker, a little chubby at various stages but with no real weight problem. She had a lot of friends. In fact, that first fall after moving to Huntington, Susan was elected class secretary, just like that. She was the girls' tennis champion at school. She had a regular golf foursome of classmates. In her senior year, when Susan tried out for the class play, *Riddle Me Riches,* she not only got a part, she got the female lead. She played a mother in that production, a role requiring her to sign a receipt on a clipboard during one scene. When, during the performance, the messenger handed her the clipboard and a serious Susan looked down to sign her name, there was a nude *Playboy* Playmate, a gift from her offstage classmates that broke her up and sent ripples of laughter through the uncomprehending audience. At the time it seemed funnier to parents than to Susan, but over the years the embarrassment subsided enough so that later she comfortably sought roles in a local production near Ruthton.

The Slaters did not travel much during their summer vacations, did not travel much, period, beyond their moves down the line. So Susan grew up envisioning the world as a tidy series of Nileses and Huntingtons, friendly, comprehensible small towns with similar people, similar interests and values, and a nice country club where familiar families went for regular rounds of golf and big evening dinners on tables with nice plates and linen tablecloths. The Slaters preferred to spend the warm, sunny days of summer at a modest cottage they rented in northern New York State on Lake Erie. It was one of those places in everyone's life that is so quiet and associated with peacefulness that years later during personal turmoil, Susan instinctively, desperately, sought to spend more time at that same place again, and her parents expected her to.

Her family was and is close and demonstrative about its closeness. The men consult the women and they feel good about that and the men don't mind and then they make whatever the decision is to be made and the women go along. When difficult times came, if they ever came, the four of them would draw closer together, and they'd talk about their mutual support and feel good and stronger for it. At sad times—and there certainly weren't many of them, just an occasional elderly relative quietly passing away—the whole family would huddle together instantly to share the emotional warmth, like prairie buffalo arraying themselves in family groupings, females in the center, males with their heads and horns on the outside for protection. On happy occasions such as Christmas or birthdays, the Slaters were just as effusive and affectionate and would lavish lots of presents on one another, which each person opened in turn as the others oohed and ahhed. One day, Susan's husband would be baffled by all this; their first Christmas together, with all his new in-laws watching, he would give his wife one luxury, a string of pearls, and something practical, a lint brush. That was it.

As expected, after high school Susan went on to college. As expected, she chose Allegheny College in Meadville, Pennsylvania, far enough from Indiana to be away from home, close enough to get home for the holidays, and still along the Erie Railroad, which the Slater family rode for free.

As expected, Susan joined a proper sorority, Kappa Alpha Theta. She concentrated on English, a normal, proper course of study. She was still interested in drama, a form of pretending socially acceptable for adults. In fact, she did her senior thesis on drama, "Recurrent Sexual Themes in Tennessee Williams's Works," which she found to be, in a word, kinky.

As expected, she studied hard among the middle-class students at the middle-class school with the middle-

class values and earned middling to better grades. She
went to the football games and she dated regularly, if
not all that seriously. She just didn't see the boys as
very capable of intelligent discussion beyond campus
gossip. Her dates were always physically modest, athletic young men, slim, well under six feet, not the
bulky behemoths without visible necks who played
football in the fall and spring and wrestled in the
winter and went to Penn State and thought Heming-
way was a running back for Notre Dame. When graduation came, she went into teaching, a safe and
eminently acceptable profession for a young woman
born four months after D-Day. The only question was
where to work; those were the days when job-seekers
could choose. That decision was easily resolved when
a certain fellow she knew and wanted to know better
decided to pursue his business studies in Philadelphia.

That was how Susan Slater, three years into her
teaching career, nearing the proper middle-class mar-
rying age, and having drifted away from the business
student, came to be in the teachers' lounge in Malvern
that weekday morning in the fall of 1968. She didn't
think much of that encounter with the substitute health
and social studies teacher, although that guy certainly
was large, no doubt larger than the real Captain Bligh.
But Rudy Blythe wasn't all that interesting at first,
certainly not enough to cancel her self-scheduled Saturday night study session when he called for a date.
Susan had let her studying slide a little for a Monday
test on Milton for an extra teaching certificate. She'd
planned to devote that entire weekend to review. But
when Rudy Blythe called a second time and suggested
that she might like to go to an Eagles football game on
Sunday afternoon, she hesitated and then accepted for
reasons she really could not explain, even a lifetime
later. In a way, Blythe reminded Susan of her father.
When he entered a room, the room got smaller. He
was tall and strong and outspoken, or blunt, and Susan was drawn to that. Outspoken and demanding
herself, she could also be very annoyed by bluntness.
She and her father loved each other very much and
remained very close. But they had had some pretty

good disagreements over the years. Father and daughter were too much alike, the railroad man would say, and Susan would agree, although neither one determined to make any changes or accommodations.

Susan found her first date with Rudy revealing. He could be blunt and loud too. It took Susan some time to learn to read the volume gauge. When he was quiet and attentive, he felt in control, but the louder he got—and he could get mighty loud—the more frightened or frustrated he was. Not many people ever saw Rudy Blythe really angry. Those who did never forgot it, although they tried. He could be charming, of course, and a little mysterious, too. Susan couldn't tell much about his feelings at first. He didn't volunteer them. One time, however, he did say that she had come within one word of never seeing him again. Rudy said he had a personal rule. If he was interested in someone, he would ask her out—twice. If she turned him down the first time, that was all right. One rejection was understandable. Two, however, was unacceptable, he told her. He wasn't going to waste his time like that. So Susan thought she had almost got a good-bye before the hello.

The Eagles game that chilled late October Sunday was one of those nondescript matches more memorable for spectators as a social event than an athletic one. Susan was on her toes that day. She always was when meeting new people, judging them and judging how they were judging her. She probably talked too much; she does when she gets nervous. Rudy talked a little loudly. At one point, seemingly disgusted by the raucous behavior of some drunken fans sitting several rows above in the stadium, Rudy stood up and gave them a loud lecture, which prompted some more shouting and a few gestures on both sides. Susan was mortified. Then Rudy sat down and offered her a drink from his hip flask.

For dinner they stopped by the house of Rudy's brother George. Everyone made his own sandwiches. Since neither Rudy nor Susan seemed in a hurry to end the day, the pair drove around Philadelphia together and discovered they shared an interest in view-

ing nice homes. Susan noted then, not for the last
time, that Rudy's taste in houses leaned much more to
the big and even massive, while hers was more toward
the tiny, cozy ones. Sometimes she thought Rudy's
ideal house would be a huge new one painted plaid
with an assumable 6 percent mortgage.

Their second date was a rugby game. Rudy starred
and Susan got bonked on the head by a ball. They
dated pretty regularly after that, and their companionship grew increasingly intense. They taught during the
day. Rudy went to his evening economics classes. From
ten o'clock until long into the night, they would sit and
talk in Susan's apartment about their lives before and
their values then, at least as they saw them and wanted
them seen. The subject of marriage in general came
up. Rudy said marriage was something a couple works
out over time. It wasn't an idyllic state they fell into
upon leaving the altar. You grew into marriage, he
said. Take his grandparents. They had seen each other
only twice before their wedding day; their marriage
and family obviously worked out. You grow into intimacy in marriage, he said, with all the wisdom of
never having done that. It was a commitment, not
always a rosy life, but a commitment. Rudy and Susan
agreed on the commitment part. They went around a
little in those late-night talks on which comes first,
intimacy or the marriage. Rudy was adamant. He'd
get that way sometimes, and no thing or no one would
change his mind. It didn't take Susan long to learn
that lesson.

Actually, Susan had already made a commitment.
She'd stopped dating other men. By Susan's twenty-
fourth birthday her mother had warned her that she
was being too picky about her men friends. "Watch
out," her mother would say, "or you'll end up like
Aunt Jane." Aunt Jane was a maiden aunt who regularly baby-sat for several families, becoming almost a
part of each. She would decorate her house with the
crayoned artwork of her young friends and keep close
track of their growth as if they were her own children.
Aunt Jane also kept a photo of a handsome man on
her dresser, a mystery man she staunchly refused to

discuss. "Who is that man?" Susan would ask. Aunt Jane would shake her head. One day shortly before her death Aunt Jane suddenly regained clarity for a moment. She grabbed Susan. "Remember the photo of the man you were always asking about?" she said, speaking slowly so every word was laden with meaning. "Well, he was married. He could never be mine. Don't you ever let yourself be like me— lonely and old and all alone."

Susan remembered Aunt Jane and her lesson and her mother's warning, even years later when the specter of loneliness loomed large. When she met Rudy, he seemed different from the others, not so low-key, more effusive. She decided almost immediately that he would be her husband. She also made it immediately, directly, and naïvely clear, although she hadn't been asked or approached directly by Rudy, that she would not sleep with him until after marriage. He replied that he'd figured that out. Since he was beginning to know Susan's determination well enough by then, too, he seemed to accept that.

There were threatening signs of change and turmoil on television every day in the late 1960s, new ways bubbling up from below that put in deep doubt old ways and standards. Rudy had felt it too. His beard, for instance. He liked it, but it was also a minor statement of freedom in those days. It wouldn't have been allowed at Haverford a decade before. Susan felt she could talk a lot more easily with Rudy than the other men she had dated. He was serious, not self-conscious, more experienced.

Like all boys, or men, he had his games. Don't they all? Take rugby, for instance. Standing out there all that time and knocking each other down in the dirt. Or hunting or fishing. Why hunt it and kill it and then have to clean it, when you can just take it out of a package from the store? But Rudy wasn't just one of those jocks. He had other interests— politics, money— and he wasn't threatened by a wife working outside the home. In fact, he was all for it; they could buy a better life then. Yes, Rudy Blythe was definitely going to have an interesting life, all right. She could just tell.

Susan, like Rudy, needed to belong to a group. From those groups she would pick out one or two women to become her very close friends. They would have long talks, sometimes daily. The conversations centered not on gossip so much as personal thoughts and events and, sometimes, fears. She could bounce a fear off a female friend and perhaps get a new insight or confirmation that her fears were okay or just feel better from talking to someone she trusted.

One time Susan caught reliable Rudy straying from the commitment she had made. They had returned from a Sunday park outing. Susan offered to make dinner, but Rudy begged off. Too much studying, he said. Got to get back to the old apartment. But late that evening when Susan telephoned to see how Rudy's work was going, she got no answer. After several tries there was still no answer, even into the wee hours. When Rudy came by the next day, Susan tried, initially in her most offhand manner, to discover what he'd been up to. He hadn't been going out with someone else, had he? Susan hadn't been, you know. Well, had he?

Rudy was not quick enough or firm enough with a denial. In fact, he didn't deny it. He talked around the question, which seemed to confirm Susan's suspicions or fears. Susan grew furious. She told him to leave, for good. It wasn't his seeing someone else, she told herself later. It was the deception. She was so angry that she slammed the door hard enough to break a window. The apartment superintendent had to come up and fix it. He didn't ask the embarrassed woman how it had happened. It wasn't the first such incident in his apartments. But it was the last time Susan ever caught Rudy lying.

It didn't take long for them to make up. One weekend shortly afterward, at Susan's suggestion, the couple drove to her parents' home near Albany as a birthday surprise for her mother. Susan remembers silently anticipating her parents' reaction to Rudy. They knew about him, of course, from late-night phone calls. Susan had always praised the man; she always would. There would be times when her father would criticize Rudy in his absence, and instantly Susan would spring to her husband's defense. Somehow accidentally on purpose Susan had, well, forgotten to mention anything to her parents about Rudy's big beard. He wasn't a hippie. Anyone could see that, so there shouldn't be any trouble, she hoped.

Mrs. Slater remembers opening her front door and, happily, seeing her daughter. Looming up behind Susan was a very large man carrying a cake box. He must be the new boyfriend, taller than the others. Bigger too. He also had the bluest eyes she had ever seen. And, oh, my God, he had a beard! What would Bill say?

Being a man, Mr. Slater didn't notice the blue eyes. And being a large person himself, he didn't notice Rudy's size at all. "The guy had a big beard," he remembers. "It was the first beard in my house. I was kind of shook up. But I didn't let on. And he turned out to be okay. A kind man. You could tell that."

Not exactly immune to the meaning of a twenty-four-year-old daughter suddenly bringing a young man home to meet her parents, Mr. Slater looked to the young man's mannerisms and how and what he chose to talk about. He was intelligent and kind and devoted to Susan, all right. But Rudy, Susan's father would come to believe, was definitely not the world's hardest worker. Over the years, Mr. Slater and his son would bustle about their house performing as a team one or another of the myriad fix-up chores they did together. Rudy would watch from a chair. He would never offer to help. If asked, he'd plunge in with energy, if not with enough skill or patience to impress a civil engineer. Rudy would come up with a course of action and try it immediately. If it didn't work, which it often didn't, he would do the same thing again and maybe again, perhaps trying physically to force the issue. Then the frustration and the impatience would mount. Anger could emerge quickly if Susan didn't step in. Mrs. Slater remembers several Christmas Eves when, late at night, Rudy could be observed sitting in the middle of the living room floor trying unsuccessfully to assemble some gift for his son. "After a while," Mrs. Slater says, "he would just give up and sit there looking forlorn until Susan or someone came along to help."

Rudy never really formally proposed to Susan. He was not the most romantic man in the world, and if he had been, Susan's bluntness, perhaps inadvertently, might have bruised such a fragile, sensitive persona. Gradually Susan and Rudy's talk of marriage just grew more prominent and both agreed, at some unrecorded moment, that it was a good idea and would be fun. So, taking a deep breath, Rudy picked up the phone in Susan's apartment one Sunday afternoon soon after meeting Susan's parents and dialed their home.

"Hello, Mr. Slater," said Rudy, nodding to an anxious Susan sitting nearby. "This is Rudy Blythe. I'm calling to ask permission to marry your daughter and I . . . pardon me, sir? Blythe, Rudy Blythe. B-L-Y-T-H-E. I was up there to visit you two weeks ago?"

They had a short, businesslike conversation. Seconds later, Susan's phone rang. It was her father. "Tell me more about this young man Blythe," he demanded. "How in the world can he support you when he's in school?"

A good question from a railroad civil engineer, and an easy answer for the son of a research scientist. Rudy promptly sent him a copy of his personal investment portfolio. Young Rudy hadn't earned it, of course. His mother gave it to him. When his grandmother died, she left a modest inheritance. Mrs. Blythe divided the funds in half and invested them in stocks for the boys. The holdings had done quite nicely, and the future father-in-law was reassured.

Susan wrote her fiancé a letter vowing eternal love the night before he became a husband. From now on, she wrote, "we'll never be totally separated." She said she could never live without him. "For the first time in my life I've met someone who meant more to me than myself," she said.

On August 9, 1969, Rudy and Susan became man and wife, a happy event with lots of friends chronicled on the back of a newspaper page that reported a bizarre cult murder in the California house of an actress named Sharon Tate. As a wedding gift, Susan's new mother-in-law gave them a week at a resort in the Adirondack Mountains, then they left with the Philadelphia Rugby Club on a two-week trip to England. Rudy played rugby and they both socialized with his friends. They toured the countryside—Rudy was especially keen on that part—and stayed in cozy little bed-and-breakfasts. In the evenings they might visit a neighborhood pub. There, sitting in the corner one night, Rudy confided to his new bride his fondest hope: to take her and return as a businessman to Ethiopia. It was a fascinating place, really beautiful. Some cities and lots of rural areas. Father Hugo was still there. The country had good opportunities. In fact, Rudy had already mailed out more than a hundred résumés to companies doing business there, seeking a job.

Susan was stunned and shocked. Ethiopia? Ethiopia was not part of the deal! She thought she had married a proper Philadelphia man and would spend the rest of her life in that bright, big city in the East, with lots of children who would go off to school so she could do Junior League work, and they'd be home in the middle of the afternoon in time for a nap before Daddy arrived from the office downtown and everyone sat down to dinner together. Tears began rolling down Susan's cheeks. Sobs seeped out. The other people in the pub, all men, were sneaking glances at the Yanks. Rudy tried quietly, and then desperately, to sell the advantages of that African land, the clean air, the fascinating culture, the countryside, the luxurious life of a foreign businessman and his wife, a car and driver and maid, everything. But the images that so fired his imagination and enthusiasm were foreign to Susan. She wasn't sold. She wasn't really listening. Her reply was basic: "Not me, buster."

That was the last time Rudy and Susan ever talked about his Ethiopian hopes. They settled down instead a world away in suburban Philadelphia in half of a duplex. They both commuted to work and school and devoted countless free hours to redecorating the aging half-a-home. Rudy still mentioned the Ethiopian idea to his friends, but it became less of a plan and more of a dream. If he ever did get any responses to his job applications, Rudy never told anyone. He began drifting from his graduate-degree work, which he never would finish, and directed his job-hunting comparatively closer to home. A friend who had been in their wedding party was working at a big midwestern bank in Minneapolis.

Rudy knew a little about the area from his university days in Iowa. Historically, Minnesota was somewhat above the main American east-ward path across the Midwest. Politically, the state tended to be more liberal, with an abiding faith in activist government, contributing one of its sons to the Democratic party ticket in every presidential election but one for two stormy decades. The state's harsh climate attracted hardy Scandinavian types, innovative, independent ethnic folk who could work the land or foster a high-tech corporate revolution. Minneapolis, reinforced by a fraternal local legion of old, moneyed families, was en route to becoming the region's nicest city, a rebuilt regional business hub, an economic and transport center with theaters, urban lakes, and shiny new skyscrapers, many of them financed by energetic Canadian money flowing down from the north.

To Susan, Minneapolis certainly was better than Ethiopia, even if it wasn't Philadelphia. She could do some teaching and some of the social volunteer work she found so rewarding, and develop a circle of close friends. Rudy was happy in his work as an economist. He and his wife could get out into the countryside easily enough, and Rudy and Susan quickly started acquiring some small rural tracts near a lake in northern Minnesota where they would go some weekends and each summer vacation.

Soon, Rudolph Hamma Blythe III was born. Better known as Rolph, the chubby little fellow was the first of several children Rudy and Susan planned. They were doting parents. He was, for one thing, proof of that love they expected to grow in marriage. Rudy had great plans as a father, and not just for Saturdays. The boy would play football, of course, and backyard baseball. Father and son would wrestle a lot on the living room floor. They would go fishing together and canoeing. They would have regular nights out together. They'd drive bumper cars, or Rudy would get Rolph to go down on the floor of the basketball court to meet the Harlem Globetrotters, and Rudy would photograph the whole scene. On the way home he'd take pictures of Rolph eating pistachio ice cream at Howard Johnson's. Every time Rudy went on a business trip alone, he would be sure to bring a nice gift home for his son, and maybe some little thing for his wife, too.

Friends from Rudy's earlier years noticed a change in their old classmate. He had become more sedate, more serious. He applied himself at work. One executive was so impressed that he took Rudy from economic forecasting in the bond department and brought him into the trust division, which seemed like a promotion at the time. Privately, Rudy began to register unhappiness there almost immediately. It was a more hierarchical department. Rudy didn't like working directly under someone, especially someone he didn't like all that much. He found the bureaucracy suffocating. He did not like the way some of the bank's trust people advised clients about what could be done with their money, and he didn't like the heavy responsibility of investing other people's hard-earned savings. He began talking about buying a farm. He'd be his own boss there. He could hire someone to help, to work when he didn't want to. They'd be out in the country all the time, all that fresh air and freedom. What a wholesome place to raise a family! He could have that horse he always wanted, and a dog too, of course. Or better yet, maybe he'd buy a bank in some small town. He could hire someone to help do the things he couldn't or didn't want to do.

Rudy began to shop around for a little bank, though none of them seemed to be quite right. Meanwhile, Rudy's dislike for his work in the bank's trust department increased. He sent out résumés. Then, one day in 1973, he announced to Susan he was off to Des Moines for a job interview. He was hired to help run some bank investments by a statewide chain of banks directed from a Des Moines headquarters.

As was to become the pattern, Susan did the initial heavy house-hunting in the new location. After a few days of touring with real estate agents, she would come up with a list of candidates ranked one through three or four, her favorite cozy little place at the top, down through her least favorite, the big one without much personality. Inevitably Rudy picked the big one. He'd often justify its purchase because the financial package was more attractive, a point he knew Susan did not have the expertise to rationally dispute. Besides, he had always lived in big homes. They were more comfortable and impressive than some dinky little cottage with trellises.

The job was definitely a step up for Rudy. An assistant vice-president, he was promoted to vice-president, then to executive vice-president, and was finally made a director of an investment subsidiary. He got a nice salary. The bank paid for his membership at the prestigious Wakonda Country Club. And the Blythes made a lot of friends, "fun people," Susan recalls, "people like us."

Rudy traveled widely around the state, informing local employees of his activities and their responsibilities. Arriving in Iowa's friendly, tidy small towns as the emissary from distant Des Moines, home of the boss, was very good for Rudy's ego. He thought he'd like that respectful treatment all the time. The travels were also very good for scouting small banks for sale.

Buying a small bank was a subject Rudy would discuss with anyone anytime anywhere, even strangers at a formal dinner. It was all he wanted to do, he'd say, his dream. Every American has a dream, and owning a small-town bank had become his. "It would have to be small," he'd say with a laugh. "I'm no millionaire." To be the president. To be making his own decisions. To have the bottom-line responsibility. To be building something of his own in a cohesive small town with that warm sense of inclusion. Some of Rudy's friends gently suggested to him that there were many unseen facets to running one's own business, especially anything as complex as a bank in something as complex as a small town. They noticed that Rudy, who could be very gracious and charming one minute, would suddenly become strong-willed and stubborn the next. "When Rudy decided something was right," recalled one Iowa friend who saw him regularly on their gourmet dinner club outings, "it was very right. And when Rudy decided something was wrong, it was very wrong. Nothing and no one would change his mind." Friends decided it was very right to listen politely to Rudy's dream, and very wrong to suggest it might entail more than he envisioned.

Susan was equally outspoken. If she had an opinion, even among strangers, she didn't hide it. But she was always willing to listen to others. Animated conversations were for her a form of entertainment, and she learned things from them except when Rudy got to talking about his small bank. Then Susan grew very quiet. She didn't criticize the idea publicly, although she wondered why, if owning a small bank was such a great idea, everybody didn't want one. In private she'd ask Rudy questions and he'd throw around a lot of numbers and words like "book value." But what she heard more loudly was his enthusiasm and his animation. "He'd get like a little boy just returning from a football game that he'd just won," she remembers. "He wanted to tell you all about it and how just everything had worked and see you respond enthusiastically too. Sure, I had misgivings. But he was a banker and an investment person. You can't throw a wet blanket on your husband's dream, and I was willing to try it. I was also naïve, very naïve."

Every few weekends the three Blythes would pack up their car on River Oaks Drive and drive into the countryside to look at yet another little bank Rudy had found in a town with fewer people than some high schools Susan had taught in. Inevitably they'd eat at the local cafe, which served coffee in cups without saucers and usually was named for the first name of the owner or his wife, like Eve's Garden of Eatin'. In the fall of 1977 a friend from Minneapolis, a man who had already bought his own bank in rural Minnesota, phoned Rudy with a rumor that the little bank in some place called Ruthton was for sale. As usual, Rudy jumped to investigate. This one seemed like a real possibility. The price wasn't too bad—$521,000—and the bank's subsidiary insurance agency, if new blood were pumped into it, might just be able to produce enough in commissions to carry the loan on the bank. If only he could get the owner to negotiate seriously. These small-town guys sure moved at a different pace.

Rudy looked at the bank's books. He checked his own assets. He applied for a loan through his old employer in Minneapolis. If he and Susan sold off their vacation lots and some stocks, they could make a down payment of about 20 percent. They could borrow the rest with a floating interest rate. At one point the accountant told them they wouldn't be in any financial trouble unless interest rates passed 17 percent. They all had a good laugh over that unlikelihood.

One day during the negotiations, which went on all fall, Susan decided to meet Rolph in the car halfway home from his Des Moines kindergarten. Driving along the winding streets of her affluent neighborhood, Susan noticed a pickup truck with a couple of scruffy-looking men in it pulling away from the opposite curb. As it left, she saw in the bushes someone with the same color coat as Rolph. It *was* Rolph! The two men had offered the frightened little boy a chocolate bar and said his mother sent them to get him. As his parents from the East had instructed him long before, Rolph had run and hid. It was the last time he walked home from school alone in the city. And Susan began to think that maybe life in a nice, quiet small town would not be such a bad thing. After all, Huntington and Niles had been good places for her to grow up.

Rudy visited Ruthton several times during the slow negotiations. Sometimes it seemed Clyde Pedersen, the bank's owner, wasn't all that interested in selling, and that worried Rudy. Then came the big day, November 15, 1977. Rudy appeared at the bank a few minutes early that morning. There were more talks, last-minute details, interminable signings, and then just as they were about done, the town siren went off. The siren is a noisy symbol of midwestern life. Depending on the hour it sounds, a wailing siren can mean it is quitting time, there is a fire, a tornado is approaching, it is time all teen-agers were home, an air raid is under way, or, on this day, it is twelve o'clock noon, time for lunch. As Rudy watched, Pedersen and all the bank workers got up, donned their coats, and headed for home to eat, politely waiting for their new owner to clear the bank door before locking it. Rudy, who had just spent a half million dollars to become president of a mortgaged dream and to become an integral part of a small town's life, was left standing in the cold on Main Street all by himself, without a key. He should have remembered that, but Rudy sometimes missed early danger signals that others saw as obvious. He dismissed the incident as funny, bought his own lunch at Alene's Cafe across the street, and returned to assume ownership promptly at one o'clock. Susan, though, remembered the incident for a long time.

Rudy's life became hectic. He hadn't realized how much he had to learn. Sometimes it was frightening. He would have to get some help in there right away. Rudy wasn't prepared for the anxiety of the towns-people either. It was fun being so prominent, but being so jolly and friendly all the time could be exhausting—learning everyone's name, reassuring everyone of his good intentions, and, most importantly, maintaining a business-as-usual atmosphere. When he was there, a good number of people seemed to drop by the bank for no apparent reason. Change is usually threatening in places like Ruthton. If residents wanted change, they would leave. So Rudy had a lot of learning to do fast—names, procedures, reputations, methods, and he was studying for his insurance license, too. A lot had to be done with that insurance agency besides changing the name to Blythe. Its income would support the bank loan, and Clyde Pederson, who had his license, had agreed to stay on for only a few months.

At the same time, Rudy and Susan agreed, something would have to be done about their house, which had come with the bank. It was nice—a big backyard, close to the office, trees well cared for—but the blue color would have to go. Although it was over twenty years old, the three-bedroom ranch-style house was Ruthton's newest. It was big by modest Ruthton standards, but small by big Blythe standards. The dining room, for instance, was tiny. How could anyone entertain in such a cubbyhole? So, while Susan stayed in Des Moines a couple of extra months to try and sell the house there and to complete her volunteer commitments, the Blythes hired a builder to knock down some walls into the garage and get in a decent fifteen-by-twenty-foot dining area. Everybody in town saw the carpenter's truck parked in front of the banker's house and the unfamiliar workman grabbing lunch at the cafe.

Say, what're you doin' up to the Blythes'?

Puttin' in a formal dining room.

A what?

A big dining room.

What for?

The man says it's too small. His wife wants a bigger one.

Ain't that somethin'? They just got here and already everythin' is too small. Must be nice. If you don't like a place, well, change it, jus' like that. Didja hear that? That Blythe woman's gotta have a larger dining room. S'pose they'll invite all of us over?

Because of the money they control, bankers are always prominent people. They are especially so in small towns where there is only one bank, and when the one banker is over six feet tall; when he has strong, loud opinions; when he wears plaids; when he comes from Philadelphia; when he buys the old rail depot and has it hauled to his house for a storage shed; when he is new.

Thanks to the automobile, rural residents can now easily take their money to financial institutions in other communities. Many people in Ruthton did, especially when Clyde refused to pay interest on checking accounts. But the other banks mustn't be too far away. The customer has to visit them regularly to apply for, to maintain, and to pay off the credit that is such an important part of rural life. Most Americans borrow money to buy a house or a car or any major purchase. Farmers borrow money to earn a living. They borrow money for new equipment, to be sure, but they also need loans every year to buy that spring's seeds and fertilizer, that summer's herbicide and fuel, that year's cattle or hog feed. The harvest comes but once a year. The bills appear more frequently. A farmer's relationship with the banker determines more than whether he gets a new car; it determines whether he gets to earn a living doing the kind of work he was raised to do. The banker's patience determines how long that demand loan will last.

There have always been times when bankers were resented or envied or feared, often at the same time. As controllers of the local dollar supply, they have so much of what everybody wants more of—money. They dress better. They drive better. They live better. They play better. Perhaps they even eat better than the folks growing the food. Dealing with a banker is also one of the very few times that a farmer, by definition an independent operator, must in effect ask someone else for permission to do what he wants to do on his own land. By and large, farmers can deal with the gamble with nature—the winds, the hail, the drought, the floods. Those events are controlled by forces larger than any individual. You can buy insurance against that kind of calamity. Only a fool would stand outdoors and rage at the weather.

But a banker is different. He's human. He's vulnerable. He is fair game. Not many people in the countryside will stick up for a banker, his wife, or his kid. He's running a business too, not a charity. He's got his own payments to make. That's not all his money in the bank. He took a risk, too. No one held a gun to his head to sign that loan paper. So you can rage at a banker, at least out of his earshot, since you might need him again someday. A banker probably has one of those college educations. He's got all those numbers and computers and bewildering legal terms. Bankers are, oh, so friendly when times are good; then they can't pass out the loans fast enough. But didja ever notice, things get tough for everyone and they're a different sort? The two-faced bastards. You gotta plead for money to work hard. You work long days out there in those fields rain or shine, and he sits in that fancy new brick building, air-conditioned and all, and he decides whether you can work for the next year. He snaps his fingers, and you get the money or you don't. Is that fair? The government gives welfare money away to folks for not working. All you want to do is borrow some so you can work. Where's the sense in that? That banker is just another ordinary person like you and me. Except he's got to have a big dining room, right? What gives him the right to . . . watch it, here he comes.

Rudy Blythe entered Ruthton with the best of intentions, considerable energy and dedication, and some half-baked plans for improvements. He didn't know what he was getting into, but he was eager to learn and to help. Rudy thought he knew a lot about banking. He really knew a lot about only part of banking, and not that many local folks had so much extra money lying around that they were interested in creating a trust. But Rudy was determined to bring some big-city imagination and style to Ruthton. They would see. Everyone would see. Rudy had some big plans to boost profits, to boost local investment and help the town with employment, and to boost himself. Over time, he had decided, he would have some impact in his little corner of the world. He would start building his own modest financial empire with savvy, branch banks, and countywide clout. Everyone would benefit by having more work and more prosperity. Rudy would be at the center of it all, the inventor of this time-release financial plan.

The plans would jell over the next few years. Rudy would put them in place. They wouldn't work, but Rudy would not realize this at first. Then he would grow frustrated and frightened and impatient. He would try, desperately and sometimes noisily, to force his plans to work, to jam those parts into place, to make them mesh the right way, his way. All around him things were changing drastically, not for the better, for reasons he didn't understand. Rudy would try to do what he knew to do. Now a big man with a big voice, he would have to stand up again on the team bench and order everyone in the stadium to cheer his team more. This time the crowd would not be listening. This time there was no crowd. This was not his home field.

For years the Buffalo Ridge State Bank, like many similar institutions across the region, had been run conservatively and safely. A handful of tellers would take in local deposits. The bank would pay the depositors between 4 and 6 percent interest. Clyde would, in effect, rent the deposits to larger, distant financial institutions for their lending purposes, earning 6 to 8 percent for himself. The difference was his income to cover costs and profit. The bank would make very few local personal loans. Clyde Pedersen could have made more money that way, but that method carried more risks, involving keen judgment of each borrower's character, broad financial outlooks, and sharp local analyses. It also required a self-assurance and the confidence that times were getting better, that local borrowers would be able or willing to repay local lenders on time throughout their commitment.

Few knew it at the time, but the country's fifteen thousand banks were about to enter a revolutionary new age in the nation's financial life, a threatening time with its roots in the free-spending era of the Vietnam War, whose legacy was lingering on many financial and social fronts. The Buffalo Ridge State Bank and its city and country cousins were about to encounter deregulation, to be cut loose from many government restrictions and protections. The government would no longer tell them what they could charge or pay for everything. That would depend on their wits and their expertise in a brave new world of tough competition and survival of the fittest. Plodding would no longer be good enough. Survival would depend on how well the bank could handle or ride through an era of startling inflation when the numbers, the customs, and the expectations of past generations would be shattered by baffling new pressures that threatened everything.

Rudy had no choice but to change the bank's fundamental direction. Few bankers did. Greater risks or not, he had to earn more money because he had to pay more money for his dream. The interest rates were creeping up. If he was going to have to pay interest on checking accounts, he would have to come up with an extra $30,000 in income to cover it.

Rudy was very good at some of his work. He would be at the bank every morning before his employees started arriving at eight-thirty. At around ten he would wander across the street to mingle at the cafe during the countryside's morning coffee break. This was the part of banking Rudy liked best, socializing. He called it public relations. He felt it gave a friendly personality to a potentially impersonal banking institution, and he was very good at it. Hadn't he done all right breaking the ice as a high school student among businessmen on the train platform back in St. Davids? And among the workers out on the Philadelphia loading docks of his girl friend's father's company? Rudy represented the bank at those coffee breaks. The bank was his. He was the bank. To need the bank was to need him. To cheat the bank was to cheat him. Personally.

Rudy would tell army stories and address people by their first names and slap folks on the back. He would ask questions of the farmers and the businessmen, and they were pleased with the interest of this big man. Not many people asked for their opinions. Getting to know Mr. Blythe, er, all right, Rudy, having him getting to know them, could help when they needed a loan. He had all that money. Word was he was making more local loans. And he was receptive.

The coffee drinkers always complained, of course, about the weather and the cost of supplies, and the low price they got for their corn and beans. Rudy said he understood. But if they thought they had it tough, let him tell them about Ethiopia and farmers over there. Now that's tough. Rudy was upbeat, optimistic, sure of the future and himself among the men. After all, things were not exactly bad now. And they were going to get better, much better. They always had, right? The figures showed growth. Rudy had worked for banks up in Minneapolis and down in Des Moines. They knew what was going on. And everyone nodded and agreed, although one or two nudged boots beneath the table.

The scene would be repeated after the regular diners arrived for lunch stomping the snow off their boots, stuffing the gloves in their pockets, unzipping their jackets, and shucking the woolen hoods on their sweatshirts or taking off their baseball caps to scratch their heads before sitting down at their usual booths with their hats tipped back and ordering "the usual."

Rudy became a regular. His coffee came without being ordered. His booth was back against the side wall near the soda pop cooler with the glass doors. He always asked about the pies. Although Gary Lindahl and his wife, Debra, were slowly assuming more responsibility for the operation of the cafe and the adjoining bowling alley, where business was dwindling, the senior Lindahls were still involved. Alene still made the pies fresh every day or two and helped out at the griddle or waiting on the tables. In the back of the cafe was a pool table that earned a few dollars and brought in some customers on rainy days. In a couple of years some video games would go there. After coffee or lunch Rudy would quietly pay his bill plus those of everyone he had been talking to, some of whom said thanks. Rudy usually carried very little money. Everybody knew that. He said so. It seemed a point of pride. Rudy either had just enough cash to pay for the lunches or he wrote out a check for the bill, plus a tip.

The evenings in those first months in Ruthton were quiet ones for Rudy, who was living alone during the construction of the dining room while Susan and Rolph finished up in Des Moines. Their furniture was down there temporarily. The real estate agent said it would help sell the place, and Susan didn't like living among boxes and makeshift furnishings. Rudy was in their Ruthton home with a mattress on the floor and a couple of suitcases strewn about, a few chairs, and a table. He ate out a lot, so he was putting on more weight. He began to jog. Late at night people would see Rudy jogging around town with his new dog, a loping, pink-eyed, short-haired German terrier named Rufus.

Rudy had picked out Rufus to replace Rolph's beloved Spot, a Dalmatian who grew mean, attacking lawn furniture and threatening to do the same to passing humans. One day Spot disappeared. Rolph's parents told him the dog had gone to a farm. Actually they had the animal killed, but being from a city, they called it "putting the dog to sleep." Rufus seemed to have a sixth sense about people. He would sidle up for affection from folks he deemed decent, and indeed they usually turned out to be decent. The day came when Rufus would not leave Rolph's side. He couldn't have even if he'd wanted to because the boy clung to his dog, patting him and talking to him intensely for long periods. With some strangers, the dog would keep a wary distance and issue an occasional bark or growl. It was a Blythe family joke that Rufus was a better judge of trustworthiness and character than Rudy. They said the master ought to take the dog down to sniff over loan applicants. If the dog sidled up to a potential borrower, give the guy the loan. If Rufus barked, forget it.

One day during those first winter weeks alone Rudy heard about some local men who played basketball over at the school on occasional weeknights. It'd be fun, a little workout, meet some people, be part of a group of guys again. After dinner, he ran over to join in. But when Rudy talked with Susan on the phone later that night, he said a funny thing had happened. He'd sat on the seats a while. Then he'd shot some baskets at one end of the gym. He sat around again. Everybody said hello and waved. But no one had asked him to join their team. Rudy was puzzled. But after talking, the husband and wife agreed the men were probably shy.

Susan and Rolph arrived in Ruthton in February 1978. She was worried about life in this dump of a town, so worried that she had started smoking heavily again. "Everything in this town," she would say, "seems so, so terribly brown." Some friends in Des Moines warned Susan that small-town life was not for everyone. "You are always on display," they told her. Or, "You won't have any privacy. Everyone knows who you are and what you're doing." Susan would reply that she had lived in small towns before, well, small cities. They're all the same. Besides, Susan would say, she didn't do anything all that exciting or worthy of gossip anyway. "We're just run-of-the-mill people too."

At least to her friends, Susan vowed to give Ruthton an all-out Slater try. Like Rudy, she'd get involved in everything. And if those people were shy around local royalty like a banker and his family, well, Susan wouldn't be. It would be up to her to break the ice. First off, she'd introduce herself and learn everybody's name the way she had back in Huntington that first fall in high school when the other freshmen chose her class secretary.

"Hi," she would say time after time as she approached someone in the bank or a store, extending her hand and hoping to learn his or her name in exchange, "I'm Susan Blythe."

"Yes," the Ruthton resident would say, unsure how to reply to anyone as important as a new banker's wife and looking startled at a woman proferring her hand so boldly, "I know."

So much, Susan thought, for the subtle approach.

Every time Susan changed homes she got deeply involved in local activities. If there weren't some, she'd start some. The groups always seemed happy to have an enthusiastic new member. Within days of her arrival Susan had joined an evening college extension course, a magazine article-writing class whose members, all women, met once a week in Ruthton. When her classmates from around the county identified themselves, they used their husbands' names—not Betty Smith, but Betty, Fred Smith's wife—as if they belonged to their men like possessions. It was a manner of speaking and thinking that Susan found foreign, and it silenced her for a moment. In those few minutes before the teacher entered, the other women chatted around Susan with animation. A main topic of interest seemed to be the new folks in town and whether or not the banker's wife, what's-her-name from that big eastern place, Philadelphia, was ever going to move to little old Ruthton. Susan was listening idly until, with a start, she realized she was sitting in on a gossip session about herself. She made a mental note: be wary. Confidentiality among friends was crucial to Susan. So she became determined, not for the last time, to cut herself off from the Ruthton women. In other towns if one group of women was incompatible, Susan would seek out another. In little Ruthton there was but one circle of women.

A lot of things were different in Ruthton and that bothered Susan. Everyone knew what everyone was doing. It was especially easy for a banker's wife who helped sort the canceled checks. Susan knew who paid whom how much for what service. She knew when alimony checks arrived, where people shopped, even how much a new dress cost. Acquaintances expected her to talk about that, but she wouldn't. Her reluctance made Susan seem snobbish. One night a man called Rudy at home about a new car loan. The next day at a women's gathering, Susan, without thinking, addressed his wife, "Oh, tell me about your new . . ." Then she caught herself and awkwardly changed the subject. That was confidential bank business. Years later, after counseling, Susan realized that she made a lot of mistakes like that, mistakes that isolated her. "I spent too much of my life at the country club," Susan would say in hindsight.

When Rolph came home from the Ruthton school the first day, he had a thick sheaf of schoolwork, much of which struck Susan, the former teacher, as busywork. She disagreed with some of the grades. She decided to spend more time at the school and did some substitute teaching, although she was appalled at the cafeteria manners she saw during lunch hour and overheard some teachers correct mistakes with, "No, no, stupid, not that way." She found herself trying to teach the students to think for themselves, even to question her statements, but she had little success. The other teachers told the students what to think, and everyone was comfortable with that method. When even a little first-grader like Rolph sought to question ways or ideas, he quickly got in trouble. Who does he think he is?

Rudy, meanwhile, was looking for help. He wanted to hire someone to look after the bank's daily operations, especially loan-making. It was so tedious and time-consuming, and Rudy knew nothing about it. He hired a man named Jerry Ihnen from southeastern Minnesota. He didn't come with too many references, but the one he did have seemed good.

Susan knew how important Jerry would be for the bank and for Rudy. But she didn't know how Jerry's wife would take to Ruthton. If she was anything like Susan, Rudy might lose his new chief assistant as soon as he and his wife saw Ruthton's rusty blue water tower. So Susan did what she wished someone had done for her on her arrival: she organized a big welcome tea, two teas, in fact, to introduce Glenice Ihnen to the other women of Ruthton. And Susan would meet them too.

Susan sat down with the little Ruthton phone book and began calling everyone. The first half of the book was invited for Thursday, the second half for Friday. By the end of the first page, Susan had not gotten one acceptance. Each woman would have to check her calendar, meaning she would have to check with her friends to see what was going on here. A formal luncheon tea? Invited to the banker's house? Me? Some women said it would certainly be nice to come, but you know, it depended on whether or not their husbands were home for lunch that day. At one point a frustrated Susan suggested to one woman that she leave her husband's noon dinner in the crockpot and plug it in. "That's what crockpots were invented for," she said. The other woman didn't laugh.

Many of the women were secretly thrilled to be included on the list of anything as exclusive as the banker's wife's formal tea. Then, in talking with friends, they discovered that the tea was not very exclusive at all. It was very inclusive. Virtually every living female in town had been invited. They were a little disappointed.

At the appointed hour, no one arrived on time. Then, suddenly, as if they had been gathering together around the corner, all the ladies arrived at the Blythe home at the same moment in a single chattering feminine mass. The plans for casual introductions and conversation turned into chaos. Instead of meeting their new neighbors one by one and having a moment to chat as each arrived, Glenice and Susan were jammed into the living room, meeting one woman after another, two or three seconds apiece. They remembered no one exactly. For a long while the slow-moving line of town women waiting to meet the new bankers' wives wound out the front door onto Leo Street, as if they had all come for a formal audience at the Blythe residence instead of a gathering of would-be friends.

Not knowing any rural caterers and wanting things to be perfect, Susan had been up since dawn making masses of little finger sandwiches. Thinking she would honor her guests by using her finest tableware, Susan brought out and polished up her silver wedding service. Susan was too busy making sure everything was going well to talk much with her guests, but she did overhear comments in the crowd about the lack of chairs. A couple of women remarked on how darling the dinky little sandwiches were, but why didn't Susan use the bigger sliced rolls that most people used around there? They made good hearty sandwiches, not one little bite. Was there, by any chance, any more coffee? Or just this stuff, this tea here? The talk about the college graduate's showy silver pots didn't get going until later at their homes on the phones. They laughed then. But Susan didn't know it.

Susan got her party supplies at Jensen's Food Market at the far end of Ruthton's downtown block. Bank customers could park by the cafe, grab a cup of coffee, do their banking, pick up the mail, and then walk thirty seconds west to the market past the Woodrow Wilson VFW Post 506 and its dusty window sign for last winter's pancake breakfast. During the day Marv

Jensen could just bundle up some dollar bills and checks and walk thirty seconds east to make a deposit in the bank, if he banked there. But since this new fellow came to town, the bank's insurance charges had really climbed, and Marv had taken most of his business elsewhere.

The market is a tiny, dark place with creaking wooden floors covered with linoleum squares turning up at some corners. A small, white-enameled meat counter stands in the back where customers can examine a very limited selection of cuts and patiently wait while Marv rings up the bill for other customers in the front at the old cash register. When the Blythes arrived, Marv Jensen was the sole survivor of five Ruthton grocery stores. He had been in the business for thirty-four years, ever since he gave up those long nights as the depot agent for the railroad. The depot is gone now. (Would you believe the banker bought it for a garage?) Marv bought the grocery. He knew everyone hereabouts anyway. Running his own business, being his own boss and free of foremen and department chiefs, had been a dream of Marv's back in the late forties. He'd fed himself and family all these years. But it didn't seem to work anymore.

People see things on television and they want them, but you couldn't carry everything, not every kind of salad dressing and soap in that little place. So people just drove the eighteen miles down to Pipestone or the twenty-five miles up to Marshall. They probably work there anyway, if they have a job. No work around here. No point in trying hard anymore, so Marv didn't rush anywhere. Marv was trying to sell out, but who'd be dumb enough to buy a used dream?

Especially when the building was falling apart. The furnace was old. The bathroom kept flooding. For a joke Marv had put a sign on the toilet door: QUIET—GENIUS AT WORK. The faucet handle had broken off, so Marv got a big monkey wrench and tightened that sucker on the little valve and just left it there. Now anyone who wanted hot water could just turn the metal wrench that was bigger than the tiny sink. Even at its tightest though, the faucet had dripped for years, leaving a thick, dark-red iron stain on the dirty enamel right beneath the wrench. One of these days Marv would have to get Jimmy Lee back in here. Now there's another thing. Jimmy Lee Jenkins. He's a small farmer. But you can't make it just as a small farmer no more either. You gotta pick up odd jobs wherever you can to make ends meet. So Jimmy did some plumbing and cement work and general fixing for folks. Put down Marv's linoleum too.

There is one customer door at Jensen's Market. Shoppers come in, angle to the left past the freezer case down one aisle maybe seventy-five feet, turn around to the right at the end, and come back down toward the door past the cash register—a simple U-shaped traffic flow that Ruthton shoppers have been following since Before Marv. Susan wasn't accustomed to such customs. If she picked out her frozen foods first, they'd be partially thawed by the time she checked out. So Susan shopped counterclockwise, coming at everyone else head-on. They had to squeeze past each other as their little carts met in the narrow aisles. A few times people said, politely, "You're going the wrong way." Susan would think to herself, "No, you are," and she wouldn't change and saw no real need to change. How could anybody seriously care which way people pushed their shopping carts in a little grocery store? It never occurred to the townspeople to change; the flow moved that way because it moved that way. And that's where the high-voltage socket was for the freezer. It was a very minor but very symbolic thing. Whenever Susan thought about shopping at Jensen's, she thought about that stupid way of shopping. And whenever a lot of people thought about Susan Blythe, they thought of a big-city woman who always went against the small-town flow.

Then when Susan asked Marv to special-order several cartons of Virginia Slims, those women's lib cigarettes, he said he'd try. But he said he kept forgetting.

Down the street at the bank, things were going pretty well. Interest rates hadn't been going down as Rudy had expected, but with inflation he'd made a fair chunk of money on the sale of his Des Moines house.

He was feeling very good. His dream seemed to be taking shape nicely. Jerry Ihnen seemed to be working out well. Rudy liked the small bank staff. They had dubbed him the Cookie Monster because whenever they bought a bag of cookies to share during coffee breaks, Rudy would mooch most of them. He took each woman to lunch on her birthday and when Valentine's Day rolled around, they each got a red rose. One year he even sent the group a singing telegram. He was the friendliest with the women who worked the hardest. They got that message.

Rudy had also taken care of Edgar Ronning. He had formally retired but filled in as a part-time teller on busy Fridays. Rudy had been impressed with him from the first day of his bank ownership when Edgar was the only bank employee to walk up and introduce himself. Rudy also paid him to attend local senior citizens meetings and represent the bank. Edgar liked that. Rudy called it public relations.

Rudy plunged into involvement in everything local he could find—government, clubs, recreation, business groups. He told his mother it was his responsibility as *the* banker. If the banker wasn't energetic, didn't have ideas and confidence, wouldn't work hard and invest locally, then nothing would ever change in Ruthton. It would continue its downhill slide. The banker was the most important person in town, Rudy said. He was the link with the outside world, the one with the money and the imagination to bring in fresh currents and benefit everybody. Finally in his life, Rudy Blythe was really going to have some impact.

Rudy was appointed to a vacancy on the Ruthton town council. Later, he was elected in his own right. He helped finance prizes for the best-decorated bike in the Memorial Day parade, where Rolph won a special citation for his costume and decorated wagon. Rudy took his turn at the July 4th Festival, sitting precariously on a lever over a large tub of cold water. A lot of people spent a lot of money throwing baseballs at the lever trying to dunk the bank president. Rudy laughed. He laughed too when he put Rolph on the precarious lever. No one dunked the boy, so Rudy tripped the lever himself.

Rudy made regular bank donations to local clubs. When Ruthton's high school basketball team got into the state tournament, Rudy rented a special room at a Minneapolis hotel and reigned there as host to whip up civic spirit—and a drink or two as well—for any Ruthtonite who stopped by. So many people went to the tournament and the hospitality suite that, while out on local patrol back home that tournament Saturday night, Lyle Landgren, the deputy sheriff, figured he was the only living soul left in town. It was unusual to see everyone so united.

When Catherine Ness and Sharon Fadness needed an enticing boost for the Ruthton Girl Scout troop, Rudy quietly put up two hundred dollars to finance a train trip to Duluth. He told them his father-in-law, a railroad engineer, would never forgive him if he didn't sponsor a train trip. You know, said the big tall man, when he was little, he rode the train every day just to go to school. The Girl Scouts couldn't imagine that because the passenger trains stopped stopping at Ruthton before they were born.

Rudy had other ideas, too. He founded the Lions Club to boost civic spirit. He suggested the town council give a break on the water rates to a local man who might bring his bait business to town. It would mean a couple of jobs and might get a little economic momentum going. Then Rudy started talking about a new investment. If he and a few local men with a little money could get a good piece of land, they could apply for a grant and maybe build a senior citizens home right there in Ruthton. There were more old folks all the time scattered around town. This way they could be together without going outdoors in the winter. They'd get good care and meals. They would stay in town and maybe their kids would, too, and help support local businesses. And Ruthton needed a doctor again, so Rudy negotiated with a doctor from Pipestone. One Saturday, he got Lions Club members to fix up a little office for the physician. Susan got three women to paint and make curtains. The club and the bank underwrote the rent for a while. And a doctor actually started taking appointments, and keeping them in Ruthton again.

In phone conversations with his distant friends, Rudy was ebullient in those first couple of years. When Rudy's father came to visit, his son would take him across the street to the cafe, and Rudy would hold court with the farmers and they'd talk about crops and the weather and the price of farm equipment. Rudy's father was impressed, and that pleased Rudy. Men would come up to Rudy on the street, they'd laugh at Rudy's jokes and ask after Susan and say it was their turn to buy coffee at the cafe. A few days later they'd drop by the bank to see their pal Rudy about a loan. There wouldn't be much of a problem. Times were good. Land was great collateral; its value was rising every day, thanks to inflation.

Some new customers even came from another bank. One, that Jim Jenkins fellow, the handyman who fixed Marv Jensen's plumbing, who lived on the road to Tyler, consolidated a couple of loans on his little ten-acre place and brought the business to Rudy's bank. With the interest rates inching up, people were shopping around more for banking services. Jenkins and the others were making their payments. Things were going swell for Rudy. The guys in the Ruthton Volunteer Fire Department, a major male social institution in most midwestern communities, even invited Rudy to become a full member. They really seemed to want him. Rudy was thrilled, couldn't wait to tell Susan. "Gimme a break," she said.

The Blythes had joined the country club in Pipestone too. Rudy got into a regular foursome of banker golfers. Rudy didn't care if the Mother's Day buffet was served on paper plates, though his wife certainly did. Generally though, Susan seemed satisfied. She was, however, putting on a lot of weight. She had tried to organize a club, like Rudy had done with the Lions, only this one would be for women to exercise together to shed some of that extra weight from all those potatoes and other starchy foods. The women would meet at the school early in the morning and run inside the gym and then have coffee somewhere together and talk. But the principal told Susan he couldn't open the school early. If he did it for her, he'd have to do it for everyone. So Susan's club died from a drought of cooperation.

Rudy noticed on the phone bill that Susan was calling her mother and some friends back in Des Moines a lot. She just wouldn't quit smoking. And that was too bad about the Ruthton school board election. A longtime member had announced her retirement from the board. Susan knew it was politically hopeless to challenge almost any incumbent in rural elections. It's too much of a confrontation. It involves that dreaded word *criticism*. Susan checked with the woman, who repeated her vow to retire. So Susan announced her candidacy. She had only been in Ruthton about a year, but she had some definite ideas for needed reforms. Then it seems there was a meeting between the school board president and the retiring member. Afterward the member said she had changed her mind; she was running for reelection again. Despite all her ideas and campaign energy, Susan finished last.

As long as Rudy took his family on regular weekend trips and let Susan make all those phone calls and go back to her parents' cottage on the lake in New York every summer, things seemed to be okay. There was some concern, and some inconclusive testing, when another baby wouldn't come along as planned. They even thought about artificial insemination. It wasn't just having a baby, though. Susan wanted Rudy's baby, another living symbol of love. Susan was feeling very frustrated about that.

Rudy had given birth to a new idea at the bank. He would open a branch office, a sign of growth and new investment, the fellow from the big city putting his money where his mouth was. The new branch would be in Holland, eight miles down the road toward Pipestone, where, it was not announced in Duane's newspaper, Rudy really wanted to get a piece of the banking action for county funds. It began with a Saturday full of public relations. At noon a real stagecoach came down the main street of Holland, Minnesota.

And sitting way up there next to the driver, waving a big cowboy hat and smiling to beat the band, was, of course, Rudy Blythe. Some "holdup men" with pistols ran out into the street. Rudy tossed "gold" coins into the sparse crowd. The stagecoach stopped by the newest branch of the Buffalo Ridge State Bank, and the president of the bank jumped down to make a few remarks about the future, about Holland, and about a new age of banking service for its 160 residents. Everyone then adjourned to the park where the new Holland Branch of the Buffalo Ridge State Bank provided everyone with free Buffalo Burgers.

Susan was there, too, that day in 1979. She had become bored and frustrated with life in Ruthton and more willing to reveal her feelings to Rudy. The town didn't like any of her ideas. When she suggested building a Ruthton swimming pool—there had, after all, been one at her club in Des Moines—they said a town pool would be too dangerous; someone might drown and their relatives would sue everyone. When Susan suggested a special Friday library night so she could expose the kids to new books while their parents did some banking, they said there weren't enough chairs. Someone even commented on Susan's clothes, so she started wearing less of the bright blues and reds she liked and more navy and brown. She needed new clothes anyway; she was many pounds heavier than two years before.

Susan didn't have a close circle of friends, although Sharon Fadness was nice and gave her some tips, like not to mention her college days quite so much. Susan's mother noticed that their phone conversations were turning more negative. Susan couldn't get a full-time job. She would be taking work from a local woman who needed it more. A banker's wife working? They don't have enough money? Susan was a bank director; Rudy wanted that. If Rudy was working late again at the office or home, she'd clean up the bank after hours and the company paid her seventy-five dollars a month for that. Her spending money.

Eighteen months into his dream, Rudy seemed more preoccupied about money. Everybody was. The Blythes had planned to combine a family vacation with a trip to the bankers' convention in Hawaii, but Rudy had canceled it, saying things were a little tight at the moment. It wasn't like him. Still, he was the same old Rudes in many ways—the plaid coats and pants, even plaid swim trunks. When Rudy found something he liked to wear, he'd wear it until it fell apart. Like the wingtip shoes that stayed so shiny always, no matter what. Handy things. And his green slacks with the soaring pheasants on the legs. Pam Bush, the wife of Lee Bush, the bank's attorney, once asked them out to dinner, but only if Rudy promised not to wear his pheasant slacks. Everyone laughed. She was only half joking.

Rudy went fishing and hunting with some friends, well, with some customers from the bank. Susan had given him a left-handed, twelve-gauge shotgun, but Rudy wasn't all that keen on the killing part. She asked him once why he went, and he described it as a social occasion, getting up early and traveling with other men and then sitting with them for long hours hidden in camouflaged blinds, talking in whispers, telling stories. Rudy liked the idea of being in the woods. "But," one friend says, "you'd never want to be lost in the wilderness with him. Not the most resourceful guy."

On his return, Rudy would wrestle with Rolph—he called him Tigerboy—and, when it was bedtime, Rudy would tell his son fanciful and sometimes silly stories about a knight named Gluckajuck. Rudy claimed they were true tales, but he was smiling when he said that. Rolph could tell his father was making them up because sometimes the knight got in such predicaments that not even the president of the Buffalo Ridge State Bank could figure out a way to extract him before the bedroom light went out. After storytime, if there wasn't some town meeting to attend, Rudy would sit in his reclining chair while Susan paid bills. And he'd check his stocks in the newspaper or read one of the biographies he liked, never novels. He didn't like true-to-life television shows either, never *Hill Street Blues,* always *WKRP in Cincinnati.* He liked classical music and country tunes, especially bluegrass, a combination that Susan found strange. Rudy's tastes didn't change much, even when the notices of interest rate increases started coming more frequently from the bank in Minneapolis.

It was Susan's penchant in those days to fix things; if something wasn't working or someone didn't understand, well, let's fix it, dammit, and right now. No better time. The first time Susan saw Rudy really angry was three weeks into their marriage right after the honeymoon on a Saturday when she started getting things organized, fixing up Rudy's schedule for that day, and issuing edicts that she didn't mean really to be edicts, though they sounded that way to others.

Rudy didn't issue edicts. Quick decisions, yes, even abrupt ones, especially if he was worried or preoccupied about something else. He enjoyed the feelings of finessing someone the few times he thought he had. Rudy needed to feel in control of a situation, with an array of choices. "If you have only one option," he would say, "then you don't have any options." Once Rudy Blythe made those decisions, they stayed made. And once he made a judgment about someone or something, it hardened into cement very quickly.

Susan wanted her relationship with the town fixed. She had tried everything that worked elsewhere, but nothing took. She tried harder, but now she started to grumble about it. Much of her discontent focused on sports. Everyone took them so seriously, yelling at their own players, sometimes angrily, laughing at opponents' mistakes, taunting them, openly trying to intimidate each other. She saw an intense basketball coach pound the floor once. A lot of hostility there. There was no effort to let every child play in every game. Some like Rolph practiced every practice but rarely got in a game. That wasn't fair. Susan wanted to fix that. Other parents were concerned; they had told her. Someone else, who had grown up with the coach or men like him in places like Ruthton, might have gone to him quietly and offered some suggestions. Susan made the suggestions in a formal letter to the coach and sports authorities. Let's get it straightened out. The coach found a lot to criticize Rolph for.

At school a male teacher might slam a miscreant up against a locker now and then, just to get his attention. And it worked. But that bothered Susan, frightened her, the muzzled violence, the pent-up storm. Rolph had to survive in that social arena by day and at home had to undergo his mother's questions about school. Sometimes on the way home he would stop by the bank. With his father, and sometimes with Susan, too, he would get a hot chocolate at the cafe or play a video game or two. Rolph would then go off to try and find someone who'd play with him. Sometimes he'd go straight home. He knew his mother would ask if anything had gone wrong that day. Sometimes she'd find him in his closet, thinking. She knew it was the school problem.

Once a teacher wondered out loud to Rolph how, being the banker's son, he could make dumb mistakes. When a teacher at one parent conference said Rolph was okay in English but always seemed slightly off in his math problems, Rudy laughed and said, "Well, we like our houses built a little crooked." But the situation didn't improve. Then one night at dinner Susan corrected Rolph for holding the fork in his fist like a scoop. "You know better than that!" she said. Rolph broke into tears.

"I'm sorry," he cried. "I have to eat one way at school and another way at home and I can't always remember which is which." Like his mother and father, Rolph had few real friends in Ruthton. Everybody knew him, of course; he was the banker's son. When invited, they came to Rolph's birthday parties that Susan organized around some theme, such as "pirates." Everyone would get a pirate's costume and a cardboard sword and hat, and they would run around the yard looking for buried treasure, chocolate pieces of eight Susan had hidden there that morning. There would be ice cream and cake and, as usual, Dad would take two pieces.

But when the parents weren't around, Rolph had a hard time. His schoolmates say it was because he was so rich and they weren't; Rolph always had a quarter for the video machines. When he got a new bike, no one seemed to notice the shiny new thing; they just made fun of the training wheels. When Grandma Slater sent him a New York Yankees baseball uniform, a real one with pinstripes and everything, Rolph went running out to the diamond to show his friends. They laughed. They said everyone in Ruthton cheered for the Minnesota Twins, didn't he know that? Rolph did, but his grandma lived in the East near New York. He'd always heard a lot about that team. They were pretty good too, weren't they? Rolph learned an important lesson that day. He never wore that New York Yankees uniform again outside the house. Except one year he put it on for Halloween and pretended to others that it was a costume.

Susan felt Rolph's school problems were worsening. She talked to Rudy about them, but he had a growing number of other problems on his mind. The last thing he wanted was a fight with the local school, the banker from Pennsylvania telling the school in Minnesota that it wasn't teaching his son properly. "I don't hear Rolph complaining," Rudy would say. "You can't protect the boy all his life." On his own, Rudy might have gone to the school superintendent and said, "Look, we seem to have a problem with this teacher. What can we do about it?" But Susan never saw a different path, not until later, not until it was all too late. Rudy sensed the need for a smoother handling of problems sometimes, although he didn't know how, exactly, and even if he had, he might not always implement it. Many times under pressure his impatience would blossom too quickly for him to think.

Their visit with the teacher was a strained session. Susan was not satisfied with the teaching at the school. They had Rolph tested psychologically. He was normal. But Susan was taken with his completion of one question: A woman is————. Rolph's answer: sometimes a mother. Susan decided Rolph should change schools. They could drive him to public school in another town. It cost an extra $1,500 a year, but what's money for? We're talking basic education.

"Can you imagine what everyone will say here?" Rudy said to Susan, his voice rising. "Ruthton is good enough to put its money in the Blythe bank, but the Blythes are too good to put their son in the Ruthton school?"

Eventually Rudy went along with the idea. Having a son in school in another town was unusual, but it might calm Susan down. Maybe she had a point, and where Rolph went to class would become the least of Rudy's worries.

One day after that Susan was going down to the bank to see Rudy about something. There are two sets of front doors on the bank, making an enclosed vestibule that serves as an insulator where people stomp the snow off their boots in the winter and remove their sunglasses in summer. Susan had entered the outer bank door and was reaching for the handle of the inner one when it flew open. The little chamber was suddenly filled with a short dumpy man wearing a dirty jacket, glasses, a stocking cap perched on the back of his head, and a scowl plastered on the front. His face was flushed pink. He stormed by Susan without a greeting or an apology. Inside, Susan asked Rudy who that man was.

"Oh," he said, "that's Jim Jenkins."

It was the end of the beginning.

And the beginning of the end.

4

ACROSS THE TRACKS

James Lee Jenkins did not normally move quickly. He couldn't really, now that he was middle-aged and the tolls of various progressive ailments were starting to add up. Short and stocky, he had a very thick body and seemingly strong legs, although they tended to swell up now whenever he overworked them, say, by hefting heavy loads in a wheelbarrow. Jim was easily winded, too.

He often complained of an upset stomach whenever he ate, which was not all that regularly since Darlene left him. Like many lonely men after a divorce, Jim Jenkins had come to see meals less as a social occasion for family in a shared home and more of a simple refueling stop that interrupted work once or twice a day. Too much time to think during solitary meals. Jim would find a diner near his work and patronize it for meal after meal, day after day.

At Alene's Cafe in Ruthton, Jim would appear for lunch. Without looking, he would order the day's hot plate special, whatever it was. Elsewhere, he would order the same hamburger or the same scrambled eggs and toast. He would sit silently at the same table and shovel the stuff into his mouth until the plate was empty and mopped up with the last piece of torn bread. Since restaurants didn't take credit, Jim would have to pay cash for his meal. He never left a tip. But he'd grab a free toothpick and shuffle out the door into the dusk. The last thing anyone would see of Jim Jenkins that day, if anyone ever thought to watch, was the dirty jacket and trousers going out the door and always, always a heavy stocking cap rolled up tight and pushed way back on his round head. Some people said they'd seen a bald spot growing under the hat the few times Jim ever took it off.

Folks didn't see Jim Jenkins around much, though, not after dark anyway. It was his eyes. The eye doctor over in Marshall had told him that he had retinitis pigmentosa, which didn't mean anything to Jim until the doctor said it was incurable and usually resulted in blindness. But Jim didn't listen to everything those college graduates said anyway. Jim never could remember the name, but he knew it was bad. That was enough. It did seem to be getting worse, just like the doctor said. Already, seeing was like looking down a tunnel. Jim would turn his head this way and that to aim his tunnel vision. At night he could hardly see anything, though he still tried to drive, driving being an integral part of manhood in rural America. Several times on the way home in the evening he'd missed his own driveway and ended up in the ditch. His country neighbors would come over with a tractor to pull the pickup back onto the road again. Everybody knew those accidents weren't on account of drinking. Jim never touched the stuff, nor cigarettes. It was his eyes, Jim said. All part of his diabetes. Well, screw it. Screw the eyes, too. And screw that special diet.

Jim had agreed to try some glasses. He couldn't see without them, but he took the bifocals back right away. They didn't work, wouldn't work, couldn't work. He wore the regular glasses pretty much all the time, except when he was milking cows. You don't need glasses to find a cow's tits, not when you been doin' it all these years.

Jim was a quiet man, which is not all that unusual in the country where men spend more time listening— listening to the radio, the winds, and the machines for warnings. He never talked much; he just worked hard, mainly with his hands, not his mind. Books and pens were for city folks. And schools. Jim was the only son of a modest farmer named Clayton Jenkins, a quiet man with very firm opinions. They were good folks, neighbors say, and neighbors know. Although how Clayton and his wife, Nina, could ever make it financially, farming only eighty acres, no one could really understand. For longer than anyone could remember, Jenkins had farmed his place—soybeans and oats mostly—just outside of Florence, another little town named for another pioneer's wife, just up the road from Ruthton. Clayton never got caught up in farm fads. In fact, he seemed to resist them. He never used fertilizer, for instance. When everyone else was on the way to doubling production, pouring on expensive fertilizers and herbicides, Clayton didn't. If he didn't have the cash in hand to buy something, he didn't buy it. In an area where nearly everyone goes to some church, the Jenkinses didn't. Oh, they used to go to the Methodist church, but, Clayton told neighbors, they were all the same; all they ever wanted was money. And the Jenkins family never seemed to be rolling in that stuff.

People around Florence would look at little Nina and they'd think, now there is a devoted farmwife, even when it wasn't all that fashionable for wives to stay at home all the time and work and never defy their husbands. "Yes, dear" was their motto. People would look at her husband, that short, broad, blunt man standing there in a workshirt and pants, his suspenders bulging over a round midriff, and they would think: now there's what Jim is on his way to being in twenty years, except Clayton had lost more hair and didn't try to hide it under a purple stocking cap. They struck friends as good parents, meaning the sun rose and set on their only child.

Like most farmers, like most people really, Clayton had a temper, not so much with machines—they can't talk back or do anything except work or break down and then you just fix them—as with humans, who aren't as predictable as machines or animals. Sometimes Clayton and Jim would argue, and Nina, the only referee for a mile or so around, would step in and calm the waters. Life would then go on as before until the next time.

In private, a farmer might take his frustrations out on a door or table or maybe sometimes the wife or the kid or himself or an animal that did something particularly stubborn when the man was feeling particularly ornery. He might fling a wrench across the garage now and then if the repairs weren't going well and the thick fingers on his chilled, greasy hands weren't doing what he wanted when he wanted. But that didn't happen much, if only because there wasn't anyone else to help and he'd just have to walk over there in his heavy boots and find the wrench himself and come back and feel silly for a minute after the explosion, which the wife in the house would swear she had never heard. And he'd lie back down on the cold cement floor and get back at the repairs. Doing them himself was the price for independence. No one around to help also meant no one around to give orders. That was right nice. In fact, if anyone ever had tried to tell him how to do it, the farmer might just as like fling him across the garage.

Clayton loved the country and his work. It was hard enough, but all he knew. If he just worked hard, the bills would get paid. If things got tight, he would work a little harder, and if things got really bad, he would work harder still. The storm would blow over presently, then he could clean up the pieces and work hard some more. They'd get by. They always had.

He was pleased with his son, too, better than a daughter surely. A son can help in the fields. Like many fathers and sons in the countryside, they were together many hours every day, sharing plantings and harvests and repairs and chores. Jim learned a lot from his father, as he was supposed to. Some of it was spoken, more of it not. Jim learned so much through farm work that he decided to quit high school. He wouldn't need any education since he was going to be a farmer like his father. His father didn't make much fuss over that decision; it seemed natural enough.

In the 1960s Jim had joined the National Guard more as a means of earning a little extra money than for any military reason. But Jim still got the basic soldier's training in marching and sharpshooting, the proper shooting positions and all. He was in a transportation unit and began honing the basic mechanical skills he was taught at home. He learned to operate heavy machinery and to take it apart when something went wrong, to find out what was wrong, and then to fix it. From then on, whenever he needed some personal satisfaction or praise, he would plunge into mechanical work. "Jim is a workaholic," his friends would say, "and he can make anything work." He didn't actually go so far as to say he liked being a mechanic, fooling around with a balky diesel engine in a drafty garage or welding something broken in the middle of a field well after dark. But whenever his friends needed mechanical or welding advice, they would ask Jim. And hours or even days later, when they heard that dumb, dead engine suddenly roar to life out back, everyone would rush outside and there would be Jim, stocking cap perched on the back of his head, standing by the resuscitated machine, wiping some of the oil off his hands and maybe smiling a little at everyone's wonderment.

There's one thing everyone remembers about Jim Jenkins: he was a hard worker. By the age of twelve he was out baling hay for the neighbors at a penny a bale. And in school he was a real loner. Not one of those bright loners who walks down the high school hallway with a bouncy step, his shirt pocket stuffed with pens and pencils. Jim shuffled into class and sat down in a far corner just as the bell rang, carrying a pencil stub in his pocket and a single battered notepad covered with doodles of designs like swastikas and cars.

You'd never see Jim at a dance or a party. He never asked any girl to go. You wouldn't see Jim shooting baskets after classes either. When the school day ended, and lots of times even before, he'd just be gone. He was there in the sixth period class, and then halfway through the seventh, someone might notice that the far corner seat was empty. He wasn't different strange. He was different familiar. Lots of guys were quiet; he was more so. Jim wasn't climbing water towers with the guys and a can of paint. He wasn't playing football. "Jim was," a classmate remembers, "to himself."

You'd see Jim in town on an errand or in someone's barn, faceless but looking out from within the black welding mask in the bright sparking light of his torch, or way out in the fields alone doing good work for hire for a few days, or emerging from Marv Jensen's bathroom to get a different wrench. You might see Jim at the town fair, moving around the edge of the festivities holding a can of pop, like Quasimodo watching the market crowd from the safety of his nearby rafters. Jim would talk briefly with a few folks. He had a very strong handshake. And everyone said he could fix anything. He might even pick up a little repair business that day, although, of course, he would never appear to seek it.

Jim was a regular part of the community. Everyone seemed to recognize him but hardly anyone really knew him. Jim didn't make appointments; he just showed up, quiet-like and unannounced, and then, when his work was done, he was gone. The one time he made an appointment, many years later, Jim used a different name.

Jim may have been on his way to a repair job on that sultry August day in 1960. It may have been due to the tall corn obstructing the view. Or it may have been a bad day for Jim's eyes. Or maybe his mind was somewhere else. But in one instant on a back road, Jim's pickup truck collided with an old lady's car, head-on hard. Just like that she was dead. One second she was a living person with likes and dislikes, children and grandchildren in the midst of a routine errand on a road she knew well. The next second she was gone forever. Jim never saw the crumpled body in the crumpled dress wrapped within the crumpled vehicle because he had hit his head in the crash. Very hard. For a while they were worried about his life. Then he started to come around. After seventeen days in the hospital, the doctor said he should see a specialist, one of those tricks Jim knew doctors often pull to get more money from patients. Screw the special doctor. Nice try. Jim didn't need advice.

Jim got married the next year. No big deal. A little

late, in fact; he was twenty-four, five foot seven, 185 pounds. Marriage was the natural thing to do. Marriage meant he wouldn't have to worry anymore about women accepting him. He'd fixed that now. She was his. She was Darlene Abraham, Laura and Reuben's daughter from over by Lake Benton, sister of a policeman, and a woman whose face would gradually grow harder and harsher in the coming years as if she were living out a long, sad country music song. Darlene's life would have all the necessary elements—pickup trucks and babies and guns and rural bars and a boyfriend and death and prison, visiting the massive struture to talk to her own flesh and blood on a telephone through bulletproof glass three feet away. Darlene wouldn't need makeup to make her cheeks look sunken.

Things went okay for the Jenkins couple at first. Of course, Jim would be a farmer; his father was, so he would be too. Jim wanted to be his own boss, to work hard in the fields and the barn by himself, to rule over the animals and his kids and his wife. Jim's son—of course, they'd have a boy—would look up to him and learn from him, same as Jim did from his dad.

Jim had some problems, one of them money. He couldn't or didn't keep track of money. Sitting around indoors with a pencil and paper and maybe one of those funny little adding machines was too much like school, especially when there was so much real work to be done outdoors. He was working hard; everyone knew that. He paid most of his bills, and things would take care of themselves whether or not he was able to balance a checkbook or figure out a rate of return on his invested capital to boost his equity while covering interest charges. Jim didn't need to know fancy words and thoughts to make his life work. No one can possibly think of everything that can go wrong. He'd just try something. If it works for a while, fine. If it doesn't, try something else. When it breaks again, fix it again. Jim was the plodding kind of guy who took one step at a time. He'd spend two days fixing an old part that would cost $5.99 new. Time and time again he'd buy old machinery no one else wanted anymore and stubbornly invest himself in its repair until it worked—for a while anyway. Jim didn't have much money. But he had a lot of time to kill, a lifetime.

There might be more efficient ways of doing things. One time when he had a herd of dairy cows and needed cheap feed, for instance, Jim would drive forty or fifty miles one-way to a vegetable processing plant that gave pea silage away free just to be rid of it. Jim would drive there every day or two and haul the smelly stuff home to feed his herd because his farm was too small for adequate pasture. Some people admired the stubborn determination of a hardworking man making use of waste products to make ends meet on a ten-acre place that used to be part of a far larger farm, even if those waste products weren't all that nutritious for the animals. Other farmers, younger farmers who hadn't quit school and who maybe even went for a few years more in the Twin Cities, would see Jim drive his little truck east and return late in the afternoon with a load of free pea silage. And they thought Jim was wasting a lot of money on gas fetching that "free" livestock food. He could put that money toward buying better food to improve the cows' milk production and quality, bring in a little more income, and still give Jim two or three free afternoons a week to invest in other farm work. That's what they thought anyway. What they said to Jim was nothing. Like as not, midwesterners think free advice is worth exactly what they pay for it. Besides, those who knew Jim well enough to offer advice knew, too, that although he was never prone to accepting it in the past, he now seemed downright opposed to even hearing it. Jim looked fine, but some part of him had leaked out on that dirt road on that sad, sultry day in August 1960.

Friends said that after that day Jim seemed to get more easily angry. He was given to "wild fits of temper," one remembered. You couldn't tell when it was coming, not even his family. One moment Jim was fine, the next he was furious. The big things, the things someone might anticipate setting Jim off, didn't seem to bother him at all. Then along would come some little annoyance, the kind of thing no one would ever see as a smoking fuse, and, *boom*, Jim was livid, yelling and swinging. One day Carl Johnson got an urgent phone call for help from Clayton Jenkins. Arriving at the Jenkins farmstead, Johnson said he found shattered glass all over and a Jenkins uncle forcibly holding Jim down after a family fight. Another time, when a broken tractor stubbornly remained broken, Jim struck it with his fist. His father urged him to take it easy. Jim looked at him for a moment and struck the tractor again, harder. Jim didn't seem bothered much by his economic frustrations, so consumed was he with working hard. Some of Jim's friends from school saw how hard their fathers—mostly farmers—worked for so little income. They scouted for work at the dwindling businesses in the shrinking midwestern towns. Then they thought, for the first time in several generations, about going somewhere else to live, somewhere familiar and not too far away where things would surely be better.

Gary Heldt and Jim Jenkins hung around together in high school. Gary hoped that some of Jim's mechanical skills would rub off on him like grease. Then Gary's father saw the family's restaurant business near Ruthton sputter. In the early 1970s he took the whole family and moved to northeastern Minnesota, to the sturdy old Iron Range country. New iron-ore mining processes and the traditional, insatiable hunger of the American consumer for things made of shiny metal had bolstered the Midwest's heavy steel industry for decades. Around the clock around the year, the helmeted men in the pits of the Mesabi Range dug and blasted the rich soil that was fed into the blazing hearths to the south in places called River Rouge, Lorain, Cleveland, Gary, and South Chicago. They were always bright, these blast furnaces. The country always needed new steel, and always would.

Hugo Heldt bought a bar to serve the men, who were always sweaty and always thirsty. When he got out of the military, Gary never even considered going back to the countryside. He joined his dad to work in the bar and also worked in a mine, and father and son worked together in a new career far from the fields around Ruthton.

Even farther away, there were new blast furnaces burning brightly in places like Korea and Brazil, where governments wanted United States dollars, and laborers were delighted for any work at nearly any wage. The interest rates were making it too expensive for American businesses to invest in new building. So those faceless decision makers who didn't seem to think much of or about the Midwest anymore started banking the furnaces in Lorain and Cleveland. The ships arrived less often from northern Minnesota; the ore-laden trains left from ore pellet plants less frequently; the miners had less to spend, so they didn't come by the Heldts' place so much. Hugo Heldt retired early.

"What's the point?" he said. The once eager young booster retired on Social Security and went back to the small town he knew, where he saw the growing gray ranks of friends gather at the post office around the first of every month when the government checks were due. Young Gary, who had grown to consider farming a demeaning waste of time—working seven days a week for a lifetime of what?—stayed to work in the mines when he could, which was less often. Gary's wife, Rita, decided to go to work as a beautician. It would get her out of the house, give her some satisfaction. At least those were the reasons they spoke of out loud, when Gary's twenty-six weeks of unemployment checks ran out.

"Things seem like they're getting harder all the time," Gary would say. "My son won't be a miner. There probably won't even be any of them when he grows up."

Gary saw Jim Jenkins now and then when he visited his father. A lot of the guys from high school were gone, of course, but old Jim was still around. Jim and Darlene had had a daughter, Michelle, and then along came a boy. They named him Steve. Jim seemed pretty much the same to Gary. Except Gary heard that financially his friend was going from one problem to another. Wasn't most everybody? Jim would start farming somewhere nearby. Things would go along adequately for a while, then there'd be some unpaid bills or bad checks. Maybe a minor court suit. Darlene would try to pay off the bills, and Jim would run up some more. Not on fancy clothes or anything like that. But there was all that gas for the truck Jim drove everywhere. And Jim never had enough money at one time to buy any good, new farm equipment outright. The steady string of repair bills on his old machinery was increasing too. Things was changin', as they say, and hard work alone was not quite enough. Sometimes Jim's father would have to bail his son out. Once Jim owed so much money that Clayton Jenkins had to put a new mortgage on his own property when he nearly had the old one paid off.

To bring in a little extra money, Jim did repair work. Or he hired himself out to help on another farm. At one point Jim hooked up with a local entrepreneur named Louis Taveirne. Louis was a strong, heavyset man with thick, strong hands and varied interests, hardworking and clever and a little feared for that by some. He always seemed to get what he wanted and to do well at whatever he tried, whether it was owning a country supper club or an isolated dump. For a while Jim ran the dump for Louis, checking what was brought in and covering it with dirt. It was a perfect job for Jim, physical, outdoors, with little human contact. Then Louis said some state inspectors had complained about inadequate soil covering the garbage and refuse. When Louis went after his employee, Jim got pretty sulky, then angry, and Louis had to fire him.

Then Jim decided to try something new during the day, when the cows were quietly making milk at home. He'd get some heavy machinery and do some of the county's roadwork. Louis had an old grader for sale. Jim saved his money and bought the machine for $1,600. He got it working right and went out to grade roads. But the county had its own machinery or hired contractors who didn't have to build their workday around a bunch of cows with full udders. Not too long after that, a matter of a few weeks only, Louis spotted his old grader for sale in a machinery lot. The dealer said Jim had sold it to him for $1,000.

Jim just wanted to milk cows anyway. He loved working with those animals. He'd be out in the barn before dawn and still out there after dark, feeding and milking and sometimes talking to the gentle creatures who value routine and familiar smells, sounds, and even familiar voices so highly. Jim understood because he liked those things too. Jim felt important in the barn. He was. He was Lord of the Stable. What he said went. The animals did what he wanted, usually when he wanted. If they didn't, he could hit them, and they'd get the message real quick. Simple animals were very predictable. They didn't have lawyers. They didn't have a lot of documents. They didn't talk back with clever phrases or twisted numbers on paper that showed things were one way when Jim damn well knew they were another.

Jim might grumble in the house about those fancy lawyers and tricky bankers. Everybody would listen and perhaps believe him. Jim was finding that people tended to listen to him much more when he was angry, so sometimes at home, he'd put on a little anger. And it worked. Sometimes a lot of real anger would come on without trying. That really worked. For a while.

Nobody said much about it at the time. Jim was considered "strong-minded," which doesn't seem so bad if you don't live with a strong-minded person. Later it would come out through the women, who seemed to have known all along that Jim was more like a dictator at home with his kids and a wife who came from a different generation than his mom.

With Jim working so hard, Darlene was pretty much responsible for raising Steve and Michelle, or Mickey. After the usual string of high school boyfriends calling at the house and sometimes helping their girl friend's father with the chores to earn a few parental points, Mickey would marry at a young age and move away with her husband. Steve would hang around with his male buddies, hunting without a license, talking guns or army or maybe talking girls the way young men think they should. Steve would wander by himself a good deal in the woods, or the modest clumps of trees that pass for woods in the Midwest. This was not so much to appreciate the outdoors and the bustle of natural life as to not be somewhere else. Sometimes he and a friend would run around the fields on the edge of town, spying on unsuspecting people and pretending to be junior commandos.

Steve struck some adults as a bit strange. He would smile or laugh when no one had said anything funny, as if he were listening to things other people couldn't hear. He had come to admire his father deeply. They both accepted each other without conditions and without naggings. There was joking between the men. The father got mad at times like all fathers. But there was a male acceptance there, not communicated in words. A handing of the wrench, without speaking, during some motor repairs as if the father were sure the boy knew what to do with it. Young men like that, even if they don't always know what to do with the wrench. Steve wasn't all that intrigued with solving mechanical problems. But he enjoyed being with his father at those times. Jim Jenkins was at his happiest then. So was Steve.

When things seemed to go right for Jim, he thought the hard work might work out all right after all. Jim had been working a small farm near Arco, Minnesota, about twelve miles north of Ruthton. He discovered in the mid-1970s how much his little piece of land there had increased in value due to something called inflation. So he sold it. He made several hundred dollars more an acre than he had paid. Just like that. Instant riches, like a businessman or a banker. Amazing and wonderful. Although he didn't understand precisely— and didn't really want to—what he had done so well, this inflation stuff made Jim feel successful. He put the money toward a down payment on a new farm four miles north of Ruthton on the road to Tyler. Sure it was expensive, almost three thousand dollars an acre, its value also having been inflated. It was a smaller place, just ten acres. Like many little plots of midwestern land today, the property once was the homestead focus of a family farm. Each day for generations people would fan out from the house to work the surrounding fields and come in for meals when the big

bell rang. But sons move away and old farmers die
and widows can't imagine running the old place without the old man. A lot of farms had been sold to
bigger farmers, who didn't need a second or third or
fourth house so much as they could use a wad of cash
for an interest payment. They'd break the house and
barn and a decent backyard off from the cropland and
sell it to someone who liked living in the country but
wasn't dumb enough to think he could make a living
off a ten-acre farm.

The farm by Ruthton was Jim's dream. Familiar
country. Money in land. A wife. Kids. Big fields—not
his, but big anyway, stretching to the horizon in all
directions. Neighbors who needed repairs. Dairy cows,
forty to sixty at a time. No one telling him what to do.

Jim got the place in 1977. He dealt with a farm
credit organization some miles away. He had a contract for deed, meaning he didn't own the land outright, but he would eventually as long as he kept up
the monthly payments. The credit office would have to
give him the land once it was all paid off.

Then this new banker fellow from the East came to
Ruthton. He bought the whole damn bank. The guys
at the cafe and feed store were saying that this Blythe
guy was making loans to local farmers right and left.
That meant Pedersen was gone. Good thing, because
he'd repossessed a quonset hut that Jim had bought a
while back. Blythe wouldn't know about that. So Jim
walked into the Buffalo Ridge State Bank and arranged a loan. The bank bought out the contract for
deed, and gave Jim a new mortgage on his little farm
with more money to expand and improve his herd. It
was all Jim's as long as he continued to make the
payments to Rudy's bank. So he did. And the two
men's country dreams were linked by ledger.

For several years Jim Jenkins and Rudy Blythe traveled in their own orbits four miles and two worlds
apart. Folks saw Jim occasionally at a local bar sitting
quietly at a table. He might sip from the same can for
hours while some of his friends laughed and danced
the evening away to the jukebox or to a live band with
electric guitars and enough amplifiers to support the

theory that all good bands are loud bands. Nobody remembers ever seeing Jim drunk; Jim said alcohol upset his stomach. But then, come the next day, many of the bar patrons were incapable of remembering what went on in those places much past eleven-thirty the previous night.

Rudy and Susan drove by those bars many times in their station wagon and saw the dim, dirt parking lots lined with mud-splattered pickup trucks parked at the door like animals at a feeding trough. The Blythes would be en route to the country club or a nice restaurant or a weekend away in a big city to take Rolph to the circus, because it didn't stop in places like Ruthton anymore.

In Minnesota many bars are licensed to sell liquor and beer to go. A friend from Philadelphia recalls Rudy telling the story of striding into such a place one evening to buy a bottle of champagne. Things grew quiet when he came in, like the principal walking into a boy's locker room after basketball practice. Rudy was familiar; not many men thereabouts were that big, and he probably had a tie on, too. But Rudy wasn't a regular in that establishment and never would be or could be. Most customers pretended to watch the television set. A few men along the bar might nod, thinking to themselves what they figured Rudy was thinking to himself. This was their turf, these were their peers, and no one wanted to look like they were kissing the banker's ass.

The bar had a terrible selection of wines, but Rudy selected one, and while the woman rang up the sale beneath the neon Schlitz sign, the banker looked down the bar with its torn potato chip bags and rings of moisture, its glasses with foam still clinging to the sides, and its little piles of change and dollar bills left on the counter to finance the next round. Rudy thought to himself how most of these men were behind in their loan payments to his dream. But he said nothing. He just left.

Rudy had begun to think of leaving the town too. Or, rather, Susan had suggested it. She obviously wasn't happy in Ruthton. Part of it was her fault, he knew.

She was having a lot of confrontations, even minor ones, because of her constant attempts to fix the things she found broken or needing adjustment. She was going to fix the Blythe family right out of the small-town banking business if she wasn't careful. At first, Rudy realized, he had excluded Susan from much of the bank's affairs. It was *his* dream. She was along as a member of the cast. Maybe, he thought, she'd come around if he confided in her more—women liked that these days—if he involved her more in the bank's affairs and decisions. It would give her an outlet, calm her down.

"You don't know how to handle these people," Rudy said, pleading for a change. "You're making too many waves." He agreed with her lots of times, of course. But he knew from his business that you don't always seize the castle through frontal assaults. Susan knew from her classroom work that if you don't decisively face up to problems early in the semester, you'll never be able to handle what comes later that term.

Susan was feeling very frustrated. Nothing worked. She had but one or two local friends to speak of, no Junior League, no culture, nothing to do. Just housework and making dinner for Rolph, and Rudy, when he came home for dinner. She was fat. She was not pregnant. Rolph was obviously unhappy too. Every day Rudy talked to her about the bank. At first he had rarely talked business with her. Suddenly, all he talked about was the bank, that damn bank. He wanted her down there, helping. Susan began to wonder, out loud, just who or what Rudy Blythe loved more. Just what was he really married to? He was working too hard. They hadn't had—couldn't afford—a proper vacation. All their money was tied up in that bank. Rudy was trying to do things for the town. So was Susan. What were they getting in return? Rejected, that's what. No one asked them over. They had gotten so used to being left out that on New Year's Eve Rudy and Susan went to bed well before midnight. One year, they had to greet Sharon and David Fadness in their pajamas when the couple dropped by for a late-night toast.

Whenever the Blythes would attend a potluck supper, no one waved and said, "Over here, join us." Rudy and Susan had to ask people if those two chairs there were taken, and the other diners at the big table would look down at their plates and mumble "no," as if the most powerful man in town and his wife with all the good ideas had some kind of communicable disease. How long does it take to get the Ruthton message, Rudy? How long, huh? Don't you see? This is a sick, violent place! While we're at it, is it so hard to push your chair back into the table after dinner? You always leave it sitting out in the middle of the room and then somebody else, namely me, has to push it back in. Even as the angry words formed in her mouth, Susan heard them as the kind of thing that could come back to haunt her. But that didn't, it couldn't stop the angry words. Rudy's dream? What about her dream of living in the city? How long are we going to be here?

Tensions were increasing between Rudy and Susan enough so that rumors floated around town of Blythe family frictions. Perhaps the bank's lawyer remembered Rudy's offhand remark about moving out of town with Susan or else she would get out by herself. Perhaps someone in line down at the bank saw Rudy and Susan through the glass window in his lobby office and Rudy turned away one time when Susan moved to kiss him good-bye. Rudy never liked public displays of affection, but the woman in line wouldn't know that. Perhaps someone passing by on a nice day with the windows open and the chickadees chirping in the budding willows heard voices yelling within the banker's house, a public display of private frustrations that the participants never knew had seeped out. Rolph and Rufus, the dog, wouldn't know the news was spreading either; they would be hiding in the boy's closet, off in a faraway pretend world.

There was some yelling, too, out north of town on the Tyler road, but no one outside the Jenkins house ever heard it. No one even hears neighbors' gunshots way out there. People don't walk along country roads anymore; everybody has a car or truck or three-wheel vehicle or snowmobile. It's about a half mile to the nearest neighbors from the Jenkins driveway, along the road, then up the Hartson driveway between the cornfields.

Hard times had set in for Jim Jenkins and his family there. In the winter especially it was harder to get about for feed. If Jim had a difficult time seeing dark ditches in summer, how well could he do with patches of ice in winter? Jim's cattle were often leading a hoof-to-mouth existence. For Jim's family, it was hand to mouth. Darlene was unhappy. He could tell. They'd been married nearly twenty years, about the time a couple should be getting considerably more comfortable in life. But they weren't. Things were not much different in early 1980 from the way they'd been at the beginning of their marriage, except they had two children, they had both become middle-aged, and the cost of everything they needed had doubled while the price of everything they could sell was static. Jim had a knack for fixing machines, to be sure, but most other things he touched turned to crap. Time after time, Jim would get an idea he thought was good and set about working it out in his own way. And, sure as sunset, it would collapse all around him, and on anyone else nearby. Then Jim would start over again, pushing the stone laboriously up the hill, only to lose it once more near the top for reasons Jim didn't know. Then there was his bottled anger. Things were bad in the Jenkins household, but they were about to get much worse.

In early 1980 Darlene Jenkins announced to Jim that she was leaving. She had had it. Enough was enough. As it happened, Darlene would stick around for a few more months. But the die had been cast and shattered at the same time. Not only was Jim's wife leaving him, she was leaving him for another man. He was Louis Taveirne, the broad-handed dump owner, the man who had been paying Jim to stay at the garbage site all day covering up the refuse people brought in. Louis Taveirne.

Darlene's departure was a devastating emotional blow for Jim. Many months, even years, later, he would talk of it with teeth-clenching bitterness, although by then Jim was blaming someone else. His wife's departure also meant the end of Jim's rule over a family fief. He'd lose his children. Mickey, who always was closer to her mother, would go off with Darlene. Steve was torn. At fifteen he still had to attend school. And what kind of household could a poverty-stricken dairy farmer run for a high school kid? Sure, they could visit each other regular-like. But it wouldn't be the everyday, in-and-out-the-back-door kind of thing. Steve would go off with his mother at first. He harbored hopes of somehow reuniting his folks. So, for a while anyway, the youth would tell each adult good things about the other. Their blunt reactions caused him to stop. Steve would choose to spend a lot of time with his Jenkins grandparents. Steve had a lot of time to kill then; he was going to be like his father, so he dropped out of high school. His grandparents let Steve ride his bike around the old farm and roam alone in the far woods, where he built dummies of logs and limbs with bottles for heads. Steve dressed these pretend people in spare shirts and slacks. He set them up in the fields and among the trees. And then he shot at them over and over again.

One evening LeRoy Knutson was driving along a back road testing a used car. He spotted something in the ditch. It was Steve Jenkins, unconscious. The boy had fallen on his bicycle and one end of the handlebar had been driven into his body. Knutson took the limp teenager to the Tyler Hospital, where an angry Jim Jenkins soon arrived. To some that day, the father seemed irrational, less concerned about Steve's injury, a ruptured spleen, than about his son being treated by someone from outside the area. Not everything Jim said came out clearly; he was very upset, and while the nurse stayed behind the counter, a local policeman was summoned. There was a lot of mumbling about outsiders and foreigners. Jim didn't want the doctor on duty to treat Steve. He was adamant about that. The doctor was a Filipino, one of many foreigners working in midwestern country hospitals now that few American doctors accepted such isolated working conditions, low pay, and rural life-style. Jim called him "a pineapple." Later, Jim made an angry visit to Knutson's service station, ostensibly over LeRoy having taken the boy to Tyler. The whole thing reminded Jim of that goddamned banker in Ruthton, Rudy Blythe. Jim said Blythe had tricked him while signing some documents; the banker had put a second sheet under the original, Jim said, so he had unknowingly signed away his land. None of this made any sense to the service station manager at the time, so he put it down to a worried parent's concern and promptly forgot it. A lot of farmers have harsh things to say about bankers. Few ever do anything about it.

A lot of bankers have harsh thoughts about farmers too. And a growing number were doing something about it in 1980. Rudy Blythe was among them. He was getting into some severe financial troubles himself, trying to carry the loan for the bank at a floating interest rate. Instead of falling, as Rudy had expected, the rate continued to climb well into the teens and was nearing 20 percent. Twenty percent! He had to take out another note on the bank. Although he was squeezing his insurance agency for all the income he could get, sometimes Rudy could only make the interest payments to the Minneapolis bank, thirty thousand dollars or so every ninety days. There was nothing left over to go toward the principal. He was barely treading water financially. Some days there would be two notices of interest rate increases in the same mail. The Minneapolis bank, which was finding a lot of its borrowers strained, was getting a little worried too. It wanted some more security for these loans. As Clayton Jenkins did for his boy Jim, so did Rudy Blythe, Sr., step in to financially bail out his son. He put up a considerable bloc of his stock as additional collateral.

The pressures were starting to build on Rudy. As they did in the 1930s and as they would through much of the 1980s, these financial strains began quietly fraying, strand by strand, the social fabric of trust across the countryside. The farmers began to think that the bankers were fooling them when they said the only way to survive on the modern farm was to borrow and get bigger. The bankers began to think the farmers were fooling them about their ability, or willingness, to repay. Rudy, too, began to think he had been taken by some of these guys, which was disconcerting to someone from the big city who valued being included so highly, who thought he was helping out, who thought he knew so much. Rudy's inclusion began to seem phony; he wondered if they had done it just to get the loans. Susan was sure of it.

Then came the Jenkins problem. Rudy thought later that he should have seen it coming, a realization that only intensified his anger. The Jenkins guy, it turned out, had a string of bad debts back through the 1970's—unpaid bills, a bad check here and there, a repossessed pickup or prefabricated workshed. Everybody except the new banker seemed to know that. Now, suddenly, Jenkins was walking away from that little farm north of town. Something about his wife leaving him. Big deal, thought Rudy. Wives are leaving husbands all over the country, and those guys don't just give up. It's not the bank's fault.

At least Rudy's mortgage covered the cattle and the land. The bank had that safe and secure. Then someone said Jenkins had been over at the cattle sales barn. He'd sold the mortgaged cattle, or most of them, and for cheaper slaughter prices, too, instead of waiting a few days for health tests and getting a much higher price. And he had left a bunch of the animals out at his farm starving. When Rudy went out to the place, he found that someone had ripped out all the plumbing fixtures, sinks, tubs, toilets. A body had to be very determined or very angry, or very both, to do that. Someone quoted Jenkins as saying that if he couldn't have that nice little farm, then nobody would.

Rudy was furious with Jenkins. Rudy was a stickler for honesty, and he thought he had been in control. But he wasn't. He'd been taken to the cleaners by a farmer, and an idiot farmer at that, the kind of moron who would drive all afternoon to get free pea silage, spending more on the gas than he saved on the feed. Rudy would pursue that son of a bitch wherever he went to get his thirty grand back. Rudy wasn't running a charity operation. He opened negotiations with Jenkins to resolve the debt, but before anything came of it, the farmer declared bankruptcy.

Jenkins didn't seem to care about anything anymore. He just disappeared, wandering around the Midwest looking for work, before he came up with a new idea, one that would surely work because it involved the Sunbelt, where the winters were mild and everyone had a job.

Rudy tried to track Jenkins for a while. He heard that Jim was trying to start up another farming operation over in Wisconsin. Rudy determined to go there and seize any property he could. The lawyers advised him that was impractical. So Rudy went to the Lincoln County prosecutor, Michael Cable. He was a young, part-time prosecutor, as they often are, gaining experience and contacts handling the area's few routine crimes to build up his private practice. Like Jenkins, Cable was going through a painful divorce. Rudy wanted Cable to file criminal charges against the farmer. Jenkins had stolen the cattle that belonged to the bank and sold them. To Rudy, it was clear as day. To Cable, it was a judgment call. Cable knew Jenkins was bad debt news. But how could he prove that Jenkins knew the sale was illegal? The guy did it locally; he didn't sneak off with the animals to another state. Cable knew of no successful prosecution in the state of unauthorized sale of collateral. Rudy could always slap Jenkins with a civil suit, though that would be like trying to squeeze blood from a turnip. So Cable said he wouldn't prosecute criminally.

Rudy got mad. At work, he began talking tougher to the bozo farmers, any one of whom could be trying to pull another fast one like Jenkins. Maybe that was the only language these people understood. Many folks remember him dismissing their loan applications with a quick wave of the hand. "That is a very stupid idea," he told one man. "Get the hell out of here and don't come back until you've got a better plan."

There were, in fact, a few farm sales, too. Rudy wasn't the only one cracking down on new loans. He and many other bankers were beginning to examine old loans. Some farmers sold out voluntarily or retired early. Interest rates going up while the commodity prices went down. Payments falling behind. Interest charges mounting. Property values plummeting. Better to get out now with a little than hang around trying hard and lose it all. Daddy may have been wrong, it seemed like.

The Blythes would make a day of it at those farm auctions. Rudy would work the crowd, talking with people, agreeing that it was sad about the sale but noting that some farm operations just wouldn't be able to make it. Susan would work the tally table in a shed, keeping track of who bought what for how much. Even Rolph earned some money, receiving twenty-five or fifty cents for running the sales slips from the auctioneer to Susan, and then promptly spending all his coins at the women's club refreshment table. Rudy and Susan would take their own lunches. When the noon break came and everyone left for an hour or so, the farm family and their friends would disappear into the house, leaving the banker and his family outside, uninvited.

As time went on, there were some disturbing late-night phone calls. One woman told Susan she hoped Rudy would die. It became a minor family joke. Susan would sometimes tell Rudy as he left for work to be sure and keep his head down low. Susan had recognized the woman's voice on the phone even though the caller had been drunk. Her kid had a loan at the bank, one of those motorcycle loans that was often past due on payments.

There were other anonymous callers. They yelled about the Blythes' dog or promised to shoot him dead. The callers complained that the Blythes let him run all around town, upsetting garbage cans and threatening children. As it happened, the Blythes, who had always lived in city neighborhoods where yards were fenced from each other, never let their dog run loose in town. Many days Susan would take Rufus out to the old Jenkins place and let him run there. The Blythes, or rather the bank, owned all ten acres now. The Buffalo Ridge Agency's For Sale sign out by the tilting mailbox was proof of that. The Jenkins place had become a kind of private Blythe park. Rufus would go crazy running around, picking up all the old scents of livestock, dogs, cats, rabbits, raccoons, deer. It was good for him to have such a big, empty dog run, dashing off into the woods while Susan strolled along behind with a twig or simply sat on the cement steps by the house, thinking of other times and soaking up the prairie sun as it filtered through the trees. It made her feel warm and good. But Susan didn't go in the house, which she thought was dark and cold and full of anger and bad feelings.

Sometimes Rolph would go out to the farm with her, and they would talk about school or Rolph would climb into the abandoned pickup truck—someone had shot out all the windows, which seemed strange—and the little boy would pretend to be one of the Dukes of Hazzard, like on TV, where country life was always happy and the lines between good and bad guys so clear. Rudy went, too, but not often. The Jenkins place seemed to be a symbol of something to him, something that he didn't enjoy revisiting, although he talked about it considerably. A few years later when friends, even distant ones in Texas and Pennsylvania, heard that something had happened at the Jenkins place, they would nod and say, oh, yes, they knew about the Jenkins place. Rudy had told them.

Every time he went to Tyler to see Lee Bush, the lawyer, or to the county courthouse, Rudy had to drive by the Jenkins place, and he'd see the For Sale sign and he'd think about the lost money from his dream sitting there, rotting in the prairie wind, all because of that single son of a bitch who fooled him good.

Even in the dark when Rudy and Susan drove to a Saturday night dinner out, the old Jenkins place loomed as a festering reminder. Every midwestern farmhouse has a powerful, tall yardlight that comes on automatically at dusk and burns all night until the real sun comes up. The Jenkins house was set back from the road a bit, and the driveway looped around the south side and ended in the back. The big light was there behind the house, and after dark every night, its rays would shine through the empty structure as though someone had turned on the lights in the upstairs windows. Many times Rudy and Susan would slow down as they drove by, thinking some trespasser had broken in. Susan found the sight eerie. Rudy didn't say what he thought.

Other farms were repossessed and put up for sale at a rate that accelerated across the region in the next few years, pushing down even more the value of the remaining operating farms, which prompted more banks to call in more farm loans and put more farms and machines on the market, pushing down values even further. But for Rudy Blythe, there was only one Jenkins place.

Rudy was feeling very frustrated. He had brought all the pieces of his dream together—the college education with its knowledge and ideas, the wife and son, the substantial financial backing and know-how, the life in the country, the bank in the midwestern small town. He had assembled them properly there in Ruthton. And the plan didn't work. No matter what he tried, this way or that, the damned thing wouldn't work. He had followed the instructions, but his college education was actually a minus in that society; new ideas were threatening to old ways. His wife was miserable. She was still putting on weight and so, come to think of it, was he. His initial investment had been eroded by all the added loan notes; he'd even had to move his bank loan to his old employer in Des Moines. Rudy knew a lot of people, fellows mostly, but you couldn't call them friends. The guy he hired to run the bank wasn't exactly setting the world on fire. Then came the election.

Up in the northeast corner of Pipestone County, Ruthton has long felt politically neglected. The town was properly Democratic in an area where Franklin Delano Roosevelt has yet to be displaced in memory as the ideal president. But in 1980 Rudy, long intrigued by politics and active in Republican state conventions in both Iowa and Minnesota, determined to boost Ruthton's role. Having been reelected Ruthton town councilman, Rudy set his sights on county commissioner. There was a need for a public spokesman. He would take little Ruthton's case to the rest of the county, stand up for the town, his little town, against an incumbent, some friend of Louis Taveirne's. Rudy was trounced. He couldn't have won even if he'd received every one of Ruthton's votes, which he didn't, it being a secret ballot and many opinions of Rudy Blythe having been kept secret too.

Rudy was crushed by the outcome. But there was no moaning or crying or complaining, nothing that anyone but a wife and maybe a personal secretary would notice. However, the confident bounce was gone from his step. His enthusiasms were tempered. His jovial talk was softer. Even when old Tom Widing from high school had six friends in Philadelphia get on the phone, each pretending to be a different secretary, one after the other saying, "One moment for Mr. Widing, please," and Tom finally came on the phone like the business mogul he wasn't, trying to impress his friend the successful banker, Rudy didn't laugh very hard. Rudy was weary.

It was a perfect opening for Susan. She could see the wheels turning in her husband's mind. One option was to sell the bank outright and get out of town. Another was to stay and tough it out, though Rudy had just about had enough of toughing it with Susan's unhappiness. A third was to try to sell the bank but, meanwhile, to get another job for extra income. That would ease the finanial drain on the bank and let it and the insurance agency, in effect, carry the bank's loan until times got better. Rudy would still draw his bank executive's salary. Then perhaps in a few years Rudy could come back, and maybe by then Susan would feel differently about Ruthton. And Rudy could keep his dream.

Suddenly Rudy seemed to agree with Susan; maybe they should leave Ruthton, he said. Rudy said he could put the bank up for sale, just get rid of it while the getting was good. Rudy said that meanwhile he would look for work around the area to help with the finances. Maybe in a few years, when interest rates would certainly have gone down, they could come back to run the bank in Ruthton or somewhere else, if Susan agreed.

In early 1981 Rudy told Lee Bush to offer the bank for sale. Susan heard him. Rudy's asking price of $900,000 seemed good to Susan; you always ask for more than you expect to settle for. But the asking price struck Lee as way out of line. Was Rudy really serious? That price was 30 or 35 percent more than a realistic one. The Buffalo Ridge State Bank was not a glittering Cadillac full of attractive features in the midst of an affluent population with unlimited potential to become a bigger bank with more resources. Small-town banks were not only small, which is a minus right off in an age of bigness, they were also vulnerable. Their market, once captive to geography and the isolation of rural America, was moving out from around them, leaving decaying small towns with little imagination and even less hope. Everyone who had a vehicle—and everyone did—could drive a few miles to the next town and bank there. Or bank by mail in Minneapolis and maybe get a free MasterCard or interest on their checking account. Even grocery stores were putting in little banking corners with talking TV screens and instant cash. So, given that outlook and given that price, Rudy's bank was a real Chevrolet. No one would want to buy the Buffalo Ridge State Bank. But, hey, wait a minute, thought Lee. Maybe, just maybe, that was Rudy's plan.

There was not exactly a line of eager buyers waiting in front of the Buffalo Ridge State Bank each morning. A couple of men did contact Lee and look at the bank's books. One of them was James Dwire, a local entrepreneur and friend of Louis Taveirne's. Rudy seemed to take each one seriously. He ordered Lee to draw up some contracts. But the potential buyers backed off or, more likely, Rudy found their final offers somehow wanting. For one thing, he said, they were too small. That made sense to Susan.

While she scanned the want ads in every major midwestern paper she could find at a Sioux Falls newsstand, Rudy began sending out his résumé. He started in the adjacent area. Maybe he could find a good financial job in Sioux Falls or teach a little economics over in Brookings, South Dakota. He'd still be close to the bank, just in case it didn't sell, you understand.

The weeks passed, and there were no job offers. Susan thought companies might not want someone who'd run his own business; he'd perhaps be too independent. Rudy's résumé went out from the little Ruthton post office to an ever-widening circle of banks and recruiters across the Midwest and then, naturally enough, down into the South and the Sunbelt. Everybody in the Midwest had good things to say about the Sunbelt in those days—the letters home from unemployed auto workers who left Detroit for the Texas oil boom, the newspaper stories about young families from troubled northern urban centers starting over in a sunny land of glass and steel where the future seemed bright and the ragged white stains of road salt didn't haunt the family car six months a year. There were jobs in the Sunbelt. Times were booming. Nice climate. Rudy flew to New Orleans one late summer weekend in 1981. He didn't take the bank job offered there. But it was heady stuff again to be flown into a city, picked up by a friendly executive, interviewed on his thoughts and ideas, wined and dined, and flown home. After four years of potluck suppers with people who never saw the worth of his plans, it was great to be patted on the back again. A real confidence-booster.

So Rudy was primed when the interview and then the job offer came from the big bank in Dallas. They wanted Rudy to help run their trust department, dealing with the investments of corporations. It was familiar turf for Rudy, and a mandate that promised a minimum of management above, telling him what to do all the time. They even flew Susan down. Back to civilization, she thought. Susan loved Dallas. More plastic and glitz than Philadelphia, and more lights and people.

Susan went on a three-day house-hunting expedition, making up the final list of three choices for Rudy to see. This time he picked her second choice. It had an assumable mortgage at 13 percent instead of the first choice's new mortgage at 17 percent. Rudy had learned a lesson about interest rates. And besides,

Rudy's pick had a swimming pool. On October 6, 1981, they made a final tour of the house before signing the purchase agreement. As they walked through the living room, the television news was on with a big story. In Egypt, President Anwar Sadat had been assassinated. He'd gone just outside town to a parade ground, thinking he was going to a military procession, a happy celebration full of promise for his troubled country. But from out of nowhere some men with rifles began firing. They saw their target as a symbol, and before he knew what hit him, President Sadat was full of holes. Weird things happen in foreign countries.

A few weeks later in Ruthton, on a snowy, dark Friday night, the Blythes packed up their station wagon with luggage and Rolph and Rufus, and eased slowly down the slippery road to the Sunbelt and the future. Rudy would return to Ruthton monthly for a few days of board meetings, to oversee things, and do some public relations—until the bank sold, that is. He could also check in with the bank by phone whenever he wanted. Jerry Ihnen seemed adequate for the moment to run the bank. Rudy hadn't announced the new arrangement around town. He feared some kind of panic, the banker leaving because he needed to earn more money to keep the bank. There wasn't any panic at all, just a couple of muttered good riddances.

Rudy's friends back East and elsewhere could tell he was less than enthusiastic about the move to Dallas, but their wives thought he was a super husband to do something like that just for his wife's life. That was real love, they said to their husbands. Rudy's job wasn't too bad. He didn't have quite the leeway he expected or maybe it was harder for a self-employed person to get back in harness on a larger team. The boss expected Rudy to jump up more often than Rudy thought was necessary. But Rudy was getting a good salary, and since no one had bought the Ruthton bank yet, Rudy was in frequent telephone contact with Jerry, sometimes a little too frequent for Jerry's liking. Once a month or so, Rudy would take a weekend and drive back to Minnesota by himself. It took two days to get there, but it was a legitimate business expense, so he drove the bank-owned car. That made it basically a free trip, and he saved a few dollars, which was good because the Dallas house alone was costing him over fifteen hundred dollars a month to carry. Besides, it was good to be the boss up there, even just now and then.

Susan thought she was waving good-bye an awful lot to the bank car's disappearing tailpipe. Upon her arrival in Dallas she found herself overwhelmed briefly by the choice of stores and the array of aisles within those stores where shoppers went in any direction they wanted and no one objected. She also found it disturbing that neighbors often talked about their new-model burglar alarms; they all had them except the Blythes. Susan would have to remember to start locking the house again, and Rudy couldn't leave his keys in the car anymore.

Basically Susan was thrilled to be living again in a real world of clubs, friends, shopping centers, and a brightly lit downtown where people met to talk about something other than corn and the weather. Susan was so busy settling into affluent north Dallas and meeting people like Beverly Grabenkort next door on Meadow Haven Drive that she was a month late getting out the Blythe family's Christmas letter to far-flung friends, but it was still better than the last three years in Ruthton, when she had dropped the project altogether. Too depressing and nothing of interest to say.

"Four years ago," she wrote, "you probably got a picture card with three happy people standing knee-deep in snow in front of the logo 'Buffalo Ridge State Bank' and a cheery little note saying we bought a bank and were just loving the simple life of a small town. True. However, the bloom faded from the rural rose about a year later for me, although Rudy and Rolph were still caught up in the quaint folksiness of small-town living for a while longer. Rolph joined my disillusionment the following year when he began to question why capital punishment was necessary for third graders. In all fairness to Rudy's flexibility, he never disliked living in primitiveness as much as I did, but what can you expect from someone who happily

spent four years in Ethiopia? At any rate, after three
years of tears, threats of suicide and divorce, he finally
decided the marriage and wife were worth saving, and
somewhat reluctantly told me he'd sell the bank and
move to an area where I wouldn't have to order everything from the Sears and Penney's catalogues."

Somebody else was on the move looking for new
work in the prosperous Sunbelt that year. He probably
passed within a few blocks of the Blythes' sprawling
home on one of his hitchhiking forays between the
Midwest and Texas, where the money was. In a sense
Jim Jenkins was always wandering somewhere. After
he'd lost the farm in Ruthton, he went into Marshall
for a few months to work as a diesel mechanic for a
bus company. He lived in a rented house trailer there
for a while during the divorce business and through his
personal bankruptcy; Darlene wouldn't change her
mind. She'd made that right clear. Jim didn't argue
anymore; he stayed silent.

It wasn't long before Jim became frustrated with
others telling him what to do, especially high and
mighty foremen who'd most likely never worked on a
diesel outdoors in the middle of nowhere with their
bare hands. Jim missed being out on the land. He
briefly rented a little farm to the north. Maybe he and
Steve could get started again milking cows. But the
barn burned down. That seemed a bad sign, and Jim
thought about trying a farm over in Wisconsin, a fresh
start in a fresh place. That idea didn't get very far. He
drifted down to southern Ohio for a while, loading
trucks for some manufacturing company. Steve came,
too, briefly, shuttling between his father and mother.
Then, like most everyone else in difficulty in the aging
Rustbelt of 1981, Jim Jenkins turned to the beckoning
Southwest. To Texas, to be precise.

Charles Snow looked the middle-aged man in the
eyes, as he always did when considering a new employee. Jim Jenkins responded with a firm gaze and an
equally firm handshake. Snow was impressed. The guy
was short and dumpy, near as broad as he was tall. He
was wearing one of those funny northern stocking caps

rolled up tight on the back of his head. Snow noticed the man's arms were thick and strong, his hands rough and blunt, the sign of a hard worker, outdoors a lot, too, and down on his luck as well, by the looks of it. That was good, very good; the guy would appreciate the work more.

"Got a place to live?" asked Snow.

"Nope," said Jim. Snow got to thinking. He ran the maintenance operations for the entire public school district of Brownwood, a central Texas town smack by the middle of the Lone Star State where the American flags go up by 8:00 A.M. and all sixty-one churches are filled come Sunday morning. It is a conservative, mid-sized city where fuzzy dice still hang from rearview mirrors. It has an economy based on nothing in particular and everything in general—a little oil and gas, some cotton and grains, explosives, clothing, pipes, sheep, cattle, and even a little dairying. One fifth of Brownwood's twenty thousand residents attend public school, and five fifths like to attend the high school football games, where a winning coach is so popular that he gets a free car with the job.

Snow and his band of two dozen maintenance misfits must keep up all eleven schools and their grounds, plus a minor fleet of vehicles, from a garage on the southwest edge of town. It's a chickenshit job compared to Snow's previous career, an exciting, thirty-one-year stint in the United States Army as a combat engineer building roads and soldiers to protect his country according to its leaders' orders. They were good times. No, they were great times. He got into two wars—Korea and Vietnam, in the latter serving three combat tours as a first sergeant, the highest-ranking enlisted man, destined to carry more experience than the privileged officers but never to become one. And determined to make up for that through toughness. "Yea, though I walk through the valley of the shadow of Death," says the Neanderthal-looking sergeant on the black velvet poster in Snow's tiny office, "I shall fear no evil, 'cause I'm the meanest son of a bitch in the valley."

For three decades Snow had worked with men, young

and old, watching them, judging them, conning them, training them to fear him but not the enemy, teaching them to build things and destroy things, to work together and to kill together. For a decade of that latest war in Asia, he had been in big demand. He didn't volunteer to go over there and get his ass shot off. Only a recruit or new officer would do something that dumb out of stupidity or honor, which can be the same thing in certain situations. But Snow never turned them down when they asked. They said they needed him. You're goddamned right, those dumb damn officers with their briefcases always needed a first sergeant to make things work. You know, first sergeants walk on water; they eat bullets; and they shit ice cream. Snow had a goddamned road to build across Vietnam and platoon after platoon of dumb shits to whip into shape. It was easy for the brass in Washington to turn on the sergeants to turn on the war; that's what they were trained to do and rewarded for doing—"Good work, sergeant." "Thank you, sir." But it was harder to turn off the war, to turn off the sergeant's fighting mentality. War can be awful and awful good at the same time for the pros. And peacetime was so goddamned peaceful. There wasn't anything useful to do, nothing to blow up or kill legally, no one to train to be tough, nothing to be rewarded for by the officers, nothing to protect except the sweet, sanitized memories. Sure you can build roads and dams anywhere, but there's no challenge. When you come back in the morning, no one has mined the new road section. There're no mortar rounds dropping in silently to get the heart thumping. No sniper up in that tree to ferret out by the telltale puff of faint, gray smoke from within the leaves.

So Snow had reluctantly retired from the military, quit active duty but not active life. He'd gotten a job back in his home state whipping a bunch of losers into a decent civilian platoon of workers, using his engineering talents, his military experience, and his knowledge of human behavior based on experience under pressure to re-create through bonding profanity the camaraderie he loves so much. "I'm a keen judge of character," Snow says. "I'm only wrong ninety percent of the time."

Snow had been having some trouble with nighttime vandals around his maintenance garage breaking things, stealing equipment and gas. Snow had put up some floodlights. He'd topped the tall fence with barbed wire, though he didn't mine the perimeter.

Snow knew better, of course. You don't survive in war or a military bureaucracy for three decades, twenty-seven years of it in foreign countries, without knowing your limits. And Snow had a better idea. This guy Jenkins didn't have a place to live or a penny to his name—something about his wife running off with an older man up in Minnesota and the bank foreclosing on his farm. "Everything just closed in on me," Jenkins had said. Looking around his cubicle of an office with the colored drawings of famous American battle scenes on the wall, Snow knew about feeling closed in. Instead of fighting for the fate of the free world with the young men he had personally forged into shape, Snow was stuck behind a goddamn desk with a foot-long Texas souvenir cigar on it talking with exterminators about the fucking termite problem at Central Elementary School. The exterminator wanted to know where the pipes ran from the basement boys' room; the termites would be there. Snow twisted the phone just under his chin and addressed the doorway toward his assistant's desk. "Have we got the plans for Central Elementary here?" Snow asked, resuming his phone conversation. Without speaking, a worker immediately jumped for the correct drawer in the metal filing cabinet. Five seconds passed before the room, and the telephone line, was filled with a single voice: "Goddammit," bellowed First Sergeant Snow, not bothering to move the phone mouthpiece, "I said, have we got the fuckin' plans for Central Elementary here?"

Snow knew about being closed in all right. So he made the new guy an offer. He'd give him a job, $3.35 an hour to start, eight hours a day, five days a week. In addition, he could live in the maintenance garage for free. He should just leave the lights on now and then, play the radio loud, and make himself seen outside the building at all hours. They'd give him a dog, a big German shepherd named Tina, for company. It was a good deal for Jim—no rent and lots of privacy except when the other workers invaded his sanctuary every morning. And it was a good deal for Snow—an end to vandalism losses with nothing for guards coming out of his budget.

Jim worked his ass off in the new job. Snow, a lean, slightly balding man of medium height with a round face and hard, tattooed forearms, knew the appreciation would show in the work. He liked losers. They made good men if you treated them right, trained them right. They valued routine and rules, and they got to read his voice. Snow felt important around these men. And why not? He was. He was First Fucking Sergeant, that's who he was. What he said, went. The men did what he wanted when he wanted. If they didn't, he would stride around, swear up a storm, maybe kick something over, and they'd get the message real quick. Once you'd learned them, simple losers were very predictable.

Snow never had a son, only two daughters, and both of them moved away with their own families. Snow knew then what his family had felt all those years he was away. To his men at work, however, he could still be an authoritarian daddy, standing there straight in fine physical shape, his wristwatch tightly secured to his left arm, his hands on both hips, sleeves half rolled, surveying their work, setting the standards. He'd watch them improve under his gaze and direction. He'd feel good, and though he'd never say it, they knew. When Snow was angry, he'd swear angry. When Snow was happy, he'd swear happy. The profane words were the same; the meanings were not.

Snow liked rough country, rough men, rough work, rough talk, rough play. Just for fun sometimes, Snow would cuff one of the younger workers on the ear, call him a dumb motherfucker, and put his hands up to box. "C'mon," he'd say, "try me. C'mon. I'll break your stupid fucking neck." The other workers would smile, getting the message of rough affection. The youth would pretend to box, but not too hard because there was a thinly disguised power in Snow's demeanor. The kid didn't want to embarrass the fifty-seven-year-old boss. But even more, he didn't want to get waxed by this combat sergeant who knew more than he let on. And Snow, who never seemed afraid and looked the stronger for it, and the kid, who couldn't hide his fear yet and so was weaker, would swear at each other and maybe wrestle a bit and walk off, calling each other names and smiling and feeling very good about it all.

Snow liked young men, especially troubled dropouts in their late teens. They were still teachable. Once they got to college, they got candy ass, thinking too much, watching their words, believing that hard brain work made up for hard back work. They were the kind that went to O.C.S. and became officers and then thought they gave the orders. In his spare time Snow was also a counselor at a reform school for juvenile delinquents, which is not what they're called anymore but which is what they were to him. Snow didn't talk about it much. What the hell for? One night a bunch of the boys went a little wild and broke out. The police, some guards, and Snow went out looking for them in the bush. Snow and a guard found a pack of them hiding. The leader had a knife. The guard was unsure what to do. Snow wasn't. He walked toward the leader. "Put it down, boy," he said. "C'mon, let's go home." The frightened youth lunged at Snow with the sharp blade. Without changing expession, the top sergeant, who'd fought little Chinese with longer knives than that before this kid was a gleam in his father's eye, nimbly stepped aside. He brought one fist down on the kid's right wrist, knocking the blade away. His other hand came down lightning hard on the back of the kid's neck. Just like that, the would-be killer was flat on his face in the dirt, the knife beyond his reach, and an ex-first sergeant holding him by the back of the neck like a kitten. The other escapees meekly surrendered.

The next day the guard filed departmental charges against Snow for physically abusing the kid. When the disciplinary panel convened, the boy, who could have been up for murder or assault with a deadly weapon had Snow not been so good with his hands, looked over at Snow, who could have broken his neck by adjusting the blow a little. The kid looked back at the guard and the panel. Physical abuse? he said. What physical abuse? There hadn't been any punches. Nothing at all. And he ought to know, right? The guard must have been seeing things out there. it had been very dark, you know. The charges were dismissed.

Of course, Snow adjusted his tactics depending on the individual. Some people he had to yell at just to get their attention. Some he had to yell at after he had their attention. Jim Jenkins was one of those Snow told what needed doing and then left pretty much alone. Jim would work along at his steady pace, slower than the younger men but a lot longer. At the end of eight or ten or even twelve hours, the younger guys would be tired and thinking of a little partying. Jim would be tired but still working steadily.

Jim worked better by himself, unless he had to measure things. Then Snow or Jim Perry, his grounds foreman, would send along some kid who was told to double-check Jim's measurements without Jim seeing him, which wasn't all that hard given Jim's eyesight. Usually they could find work for Jim to do alone. First, they tried him on cement. Jim struggled with that heavy, full wheelbarrow. As often as not, Jim would lose control of the wheelbarrow at some point and spill the fresh wet goo. Or he'd have to stop every few steps, huffing and puffing for air, he was in such lousy shape. Within the hour, his legs would start to swell up, but he would plug away.

Then they put Jim on grounds work and mechanical repairs, fixing mowers and trucks. Whenever the other men took breaks, Jim would keep on working or go off by himself to eat and rest. When everyone else went home to shower and play and rest, Jim would stay at work. He had no shower there; the garage bathrooms were labeled POINTERS for men and SETTERS for women. There was only one woman there, Peggy Dobbyns, the secretary, who would one day be the last one in Texas to talk to Jim.

For the first few months, while Rudy and Susan Blythe were 160 miles away shopping for their north Dallas home with the swimming pool and 13 percent assumable mortgage on Meadow Haven Drive, Jim Jenkins lived in a small wooden loft. It was on Stewart Avenue above the maintenance garage, up among the light bulbs and air filters and toilet paper cartons. It was hot up there by the roof at the top of the red, raw wooden steps. Jim slept in his clothes on the floor on a piece of foam rubber he would roll up by day. From there, he could see under the railing down into the machine shop with the broken motors and flat tires and cracked school desks. For worker reference, there was a four-year-old calendar with a color photo of a woman in a very small, gold bikini; she was holding a very large power drill and looking content to be in the viewer's presence. The radio on the workbench was one of those battered plastic cases perpetually emitting country music. No one ever turned it off; if they noticed it was still on, they just yanked the plug out.

Outside the rippled steel doors, the lot was strewn with old wheels, sinks, pipes, dead air conditioners, and other construction refuse. Across one street was a lumberyard. Across the other was a Ralston Purina feed plant. The sound of tons of cattle pellets rattling inside metal silos mingled with that of rope banging rhythmically on the hollow metal flagpole. It is a quiet semi-industrial neighborhood with a band of small houses and scruffy lawns. At night, when each facility's tall floodlights come on automatically, the area's dogs can get to barking, each rising to the vocal challenge of the other until they are all yelling. And then, one by one, they each get tired and quit, or one huge canine throat commands silence. And they all obey.

Many nights Snow would drive by the maintenance garage, looking for troublemakers. Instead he'd find Jim under a truck or fixing a tire, putting in a few extra hours of work, unpaid, just to have something worthwhile to do. He couldn't just do nothing. "I figured I'd gotten me a treasure in Jim," Snow recalled later.

Jim had floated around a few other jobs in Texas.

He worked for a traveling amusement park named
Wimpy's, helping to maintain the rides, setting them
up, and dismantling them. He was a shipping dock
worker at a manufacturing plant. In his spare time he
wandered in search of a new farm, a dairy operation,
naturally. One day Jim heard that James Lancaster,
Brownwood's school superintendent, had an abandoned
tractor at his little farm outside town. It was a John
Deere machine, a very old model with big tires chewed
by cows and parts rusted by seven years of disuse. Jim
told Mr. Lancaster he used to have one like it and he
thought he could fix it up. The superintendent laughed
and said, "Go to it," knowing full well the thing was
beyond repair. First, Jim patched the tires. Then he
towed it to the maintenance garage, and during his
long idle hours alone, he disappeared beneath the
machine.

One Saturday months later, there was a knock on
the front door of James Lancaster's house. It was Jim
Jenkins, come to make a point. There, parked out by
the curb, was the old John Deere tractor, chugging
away, its powerful voice restored. Superintendent Lancaster exclaimed at length over the minor mechanical
miracle. Jim didn't say much. He just smiled.

Snow said he could use the tractor now and then out
at his country place, which aroused Jim's interest.
Snow had two hundred acres about twenty miles west
of town. It was small by Texas standards, just a hobby
for an ex-soldier with a penchant for the countryside
and a pension from the military to invest. The land
was twenty times larger than Jim's last place. It was
large enough to graze and grow a fair crop of hay.

That year, right after Snow had cut and baled the
hay, he had to undergo a hernia operation. Jim offered to bring the bales in. Snow said, "I can't pay
ya."

Jim replied, "I didn't ask."

That weekend Jim showed up and spent two days
lifting, stacking, moving, lifting, and restacking more
than fifteen hundred bales of hay, stopping just long
enough to down a couple of Snickers bars. "My God,
he was happy out there," Snow remembers. When Jim

finished, Snow felt guilty and stuffed forty dollars in the worker's shirt pocket. Jim gave it back. "I just wanted to get back in the fields," he said.

That was the beginning of a closer relationship between Snow and Jim Jenkins. They both liked being outdoors on the land. They both liked being around animals. Snow kept a few dozen head of cattle to graze on the tough grass that struggled under the Texas sun among the squat, wiry mesquite trees. To most Texas ranchers the mesquite is a noxious weed, one that takes up valuable grass-growing space and, more important, sucks up too much precious water. So once in a while ranchers bring in a bulldozer and level a few dozen trees, pushing them together in a big pile to dry out for three or four years. Then they throw some old oil on the pile, light it, and let it burn for several days. They do this very carefully because during the drying period, the mound has usually become home for families of rattlesnakes, hiding in the shade. What looks like an innocent group of trees, local people soon learn, can suddenly become a deadly place to be.

Snow kept a few more trees than other ranchers. Soldiers, even retired ones, don't feel all that comfortable too far from cover, and mesquite made a good windbreak for the animals. They attracted a lot of birds. Snow likes birds and can point out all kinds. Like Jim, he wasn't all that fond of hunting animals. People would kid Snow about being a soldier and hunting the enemy, but not liking to hunt deer. Snow would patiently say that's right, he didn't mind hunting people who were hunting him. In fact, he'd say with an attempt at a straight face, in the army he had to kill two or three people a week just to keep his rank.

Jim and Snow and maybe Norm Cox, who helped Snow out too sometimes, would talk while they were working outdoors in the warm Texas evenings or on weekends. Jim would get to grumbling about bankers, especially one prick back home, and Norm, who also grew up on a farm, would agree they were bad folks, leading you on to borrow when you might not want to and then leaning on you to pay when you might not be able to. They can get pretty nasty. The three of them agreed things had changed a good deal in the countryside since they were young. Norm's family farm, for instance, could no longer support the father and son, so Norm had to go into upholstery and then maintenance work and help on the farm when he could. The way the banks and governments has got things organized, Norm would say, you can't hardly afford to have but one job, especially if it's farming. You can farm only and go broke for sure. Or you can farm and have another job and maybe make it. Time was, Norm recalled, when someone had to put up a new barn, the whole neighborhood got together to help on a Saturday. They made a picnic out of it. Now, Norm said, anybody's got a barn to go up, they do it themselves. And all the neighbors ask themselves where that guy got all the money to do that. Jim would go on about his home farm country, how beautiful Minnesota was, how black and thick the soil was compared to Texas sand, and how green and rich the hay was. Texas hay, Jim said, would be just straw back home in Minnesota. Jim said he liked Texas, especially the mild winters. But he was saving his money to go back up north and farm again someday. That made sense. Yankees are human too. Like southerners, they all want to go home sometime.

Jim seemed to have a way with animals, so it was logical for Snow to turn to the quiet Yankee for help. Snow had had Jim Perry shop around for a horse for one of his daughters, a living gift from the daddy who was away in a foreign country for those many years of birthday parties. Everything was going fine until Snow saw the horse lying down in the pasture for the longest time. It was sick and wouldn't, or couldn't, get up. The vet was summoned. He gave the animal a few days to get back up on her feet before she would become retarded and crippled. "If she doesn't get up soon," he said, "it'd be best to put her to sleep." For the better part of three days Jim Jenkins sat out in the pasture with that horse, talking to her, feeding her, patting her, encouraging her to get up, looking as a mechanic might for the sound or signal that would reveal the malfunctioning part. The deadline came. The horse was still down. Perry walked out into the pasture with a .22 pistol. He held it toward Jim.

"No, sir," said the farmer, "no, sir. I can't do it." And he walked away into the mesquite with his fingers in his ears.

One day about six months after Jim's arrival, a kid shows up and starts hanging around Jim all the time. The kid looks different; maybe they dress that way up in Minnesota but not in central Texas. His head is shaved. He's always wearing camouflage shirts and pants, sometimes with a big knife strapped to his leg. The kid and Jim talk together. They live together in a tiny trailer next to the garage. They eat together. They work together as if one knows what the other is thinking without saying anything. The kid is always around, even though he isn't on the school district payroll. Everybody figures the kid is Jim's son and he's on leave from the army or somewhere.

They were half right. He was Jim's boy, Steve, although no introductions were made. But he wasn't in any real army. Steve had been interested in the military since early childhood; family snapshots show him in a uniform by the fourth grade. Like many country kids, he was around guns at an early age. But his fascination with being a soldier never waned. In fact, it intensified. It was a ticket to somewhere else as someone else. Whenever Steve had a couple of dollars, he would buy adventure magazines and *Soldier of Fortune,* with grainy photos that always showed the other guy getting wasted in somebody else's jungle. Steve had set his sights on the Marines—the tough guys with the big guns that got all the respect. After the divorce Steve talked to a Marine recruiter with a shaved head, perhaps seeking another kind of family. The soldier in the creased pants and spotless shirt told the kid with the unruly black hair in the jeans and T-shirt to forget it; he wouldn't make it in the Marines on account of his missing spleen. If joining the regular army ever occurred to the teen-ager, he didn't act on it. All those TV army ads talked about training and technology anyway. The thought of going back into a classroom, any classroom, was the last thing that would appeal to him. Steve just wanted to be a soldier. What d'ya need classes for? So Steve, who as a child had painted his red wagon a khaki green with a white star and *US ARMY,* enlisted in his own private army, a one-man battalion with all the gear and weapons and lingo he could find and buy. He'd get his training by himself. His father said some guy down in Brownwood knew a lot about military stuff. He'd been a sergeant or something in Vietnam.

Steve, at seventeen, didn't care all that much about going back into dairying, although he'd do it if his faher did. All the mucking about in mud and cow shit didn't excite him. What got Steve's attention real quick was any talk about military hardware, the guns, the knives, and the stories of hard hand-to-hand combat in faraway places. It was glorious and removed from everyday humdrum routine and safely exciting at that distance. A convincing war story is unconsciously edited either by the teller, who must forget things to survive, or by the listener, who forgets the scary parts. The veteran remembers the mortar rounds and the face of that one gook when, for an instant, their wide eyes met through an open door, and neither one fired, but he forgets, or tries to, the wet on his sleeve that once was part of a friend.

What got Snow's attention was the kid's fascination with the military and soldiering. The kid talked more than his dad and had a chip on his shoulder. He often seemed ready to argue. At times, he was given to frantic outbreaks of activity, like the time Jim and Steve were assigned to help renovate an old building by removing some walls and windows. Steve went wild smashing the windows out one after the other and making quite a mess instead of prying them out one by one and carrying them to the dumpster. One day when Steve appeared at the office to pick up his father's paycheck and Perry wouldn't give it to him, Steve flew off the handle, calling the tall foreman, the man who had introduced his father to Snow, "the son of a bitch of the month." Or the time a female friend of Steve's was complaining about another woman and he offered to build a homemade time bomb by hollowing out golf balls and filling them with explosives to drop in the woman's gas tank.

Of course, Snow knew nothing of this. He had seen far worse recruits in his day. They'd come off the bus, those little peckers, trying to look tough but really looking lost, dropping things, forgetting their right from their left, closing their eyes when they pulled the trigger, freezing up when the shells whistled overhead or exploded with a crunching *whumpf* nearby. There was nothing different about this little Jenkins kid except he was a darn good listener. Snow knew his stories were good. But still, it was flattering. And Steve had been reading the manuals. Snow could tell, because one day Steve asked him about the caliber of the standard NATO rifle ammunition and Snow had said 9 millimeters and Steve came back the next day and said, no, it was 7.62 millimeters. Snow didn't say anything then, but that night he checked it in the book himself and, goddamn, the little shit was right.

The kid was full of questions that Snow liked to think about. What's it like in battle? Do you actually see the enemy? How's it feel to kill somebody? Were you ever afraid? Sometimes Steve's barrage of questions was so heavy, and aroused such intense memories in Snow, that a tear or two would start to form in the corner of his eye. God help him, he had loved it so! He had been so damned good at what he did for his country. He had done it so well. He had brought his men through, too, made each one a little more of a man, taught them they could do things they never knew. Snow had survived, which was a helluva badge in itself, given all those who hadn't, although he couldn't remember all their names. No one ever saw his tears. Snow would whip out a big handkerchief—you know, they make good tourniquets if you ever need one— and blow his nose real natural-like and accidentally wipe his eye as if he'd gotten a speck of sand in there somehow.

"You kids are all alike," he would say to his recruits, "a bunch of little shitheads that don't know fuck-all. You think you do, but you don't. You just make targets for the VC to practice on unless you smarten up and lissen to me, boy. You can't even fool your own kind."

There was the time in Vietnam the colonel said he needed Snow to shape up an American camp that was getting hit all the time and taking big losses. The colonel said he couldn't figure out what the hell to do, but he needed the first sergeant and his experience. The last thing Snow wanted to do was go back up front, wherever that was in Nam. Five minutes after he arrived, when the chopper had left, the VC suddenly attacked. Those little gooks were running all over the camp. The dumb fucking Americans, where the hell were they? This was the enemy that was running all over camp. They were shooting machine guns and running between tents and buildings lobbing grenades and satchel charges in where the Americans were hiding. Hiding? *Hiding?* Well, this was Snow's fucking camp now. Goddamn son of a bitch! If any fucker was gonna be hiding in this camp, he'd be hiding from Snow, not from these little shits in shorts. Jesus H. Christ, things were gonna change around here. And it was gonna start changing in about two minutes. Now get everybody together, goddammit. *Now!*

First things first. Snow announced a new policy on inspections. They were gonna have them real regular. There was a collective groan from the assembled men, which was followed by the sound of a first sergeant whirling around to aim his eyes at the group. Did someone say something? It seemed not. There was silence in the camp as the message began to sink in. Now, I want this place cleaned up. This is a goddamned pigsty! Maybe you assholes live in a pigsty back home on the farm. Well, Snow didn't. And Snow wouldn't. His goddamned camp was not gonna be a fucking dump. Didja hear him? In about fifty-four fucking minutes Snow was gonna walk down this same path and things had better be in better shape.

In fifty-four fucking minutes precisely, Snow walked

down that same path and what he saw appalled him. He still saw a goddamn mess. Maybe it was even worse than before. He'd never seen such a mess. There was something wrong everywhere. He saw things wrong that the last first sergeant had seen as right. Snow wanted things to shape up. *Shape up!* Unnerstan? Now, goddammit. *Now!*

Snow instituted early-morning inspections followed by morning inspections followed by midmorning inspections and late-morning inspections, noon inspections, lunch inspections, midafternoon inspections, late-afternoon inspections, dinner inspections, and evening inspections. In a few days Snow had something good to say. The camp was starting to look barely decent, he said. As a reward he was dropping one inspection. Instead of being inspected beneath the hot afternoon sun, his men were going to practice repelling an enemy assault in the hot afternoon sun. Hand-to-hand. Rifle butts. And target practice out in the open where everyone, anyone, could see and maybe think about it, in case they planned an unexpected nighttime visit. There were new foxholes dug and lots of shooting practice. That's what the fucking rifles were for—to shoot the enemy—not to throw down or hide behind. Teach these pansy asses how to defend themselves. Maybe these dumb bastards wanted to die cowering in their tents. Snow didn't. Snow was not an officer. He was better than that. In this camp he was—how should he put it? —God. Yes, God. That had a nice ring. But the men could call him First Sergeant for short.

Oh, and the mines. Snow wanted more of them out on the perimeter. If they were gonna have visitors, it was only right that they be greeted properly. Snow liked mines. They were music to his ears, going off in the darkness with their sudden, distinctive crash.

Things started to change in Snow's camp. People dressed neater, acted neater, thought neater, even without all the inspections. When the VC started lobbing in their mortar rounds in the darkness, Snow's men didn't run and hide anymore. They stayed down in their new foxholes. Maybe they were more frightened of the first sergeant than the mortar shells, but when the enemy mortar rounds stopped falling, they started shooting just like the first sergeant said he'd be doing.

One night the VC attacked with special force. Snow made a last-minute check of his troops up front, striding along in the heat, shirtless, carrying nothing but a stick, talking tough. Then the mortar rounds came sailing in. The radio operator wanted to know if maybe perhaps First Sergeant Snow wanted to call in the gunships now.

"What for?" asked the sergeant.

They kept up the mortar barrage longer than usual, Snow noted, which was a tip. So he had his men start firing into the darkness even during the mortar barrage. And while the guys were slamming new clips into their weapons, and between the mortar blasts, Snow's men could hear the mines going off here and there. Those little fuckers were trying to sneak men into Snow's camp even amid their own exploding mortar barrage. How had the sergeant known?

The radio operator wanted to know if maybe now was the time the first sergeant would like him to call in the gunships. The first sergeant said he was a little busy to chat with the radio operator right then. With all the guns and mines and shells going off, the last thing Snow wanted was a bunch of helicopters *whop-whopping* all over the place, shooting up his camp and making all that noise. Besides, he didn't want to disturb the pilots' sleep.

When the heavy mortar barrage stopped, the assault came. A mine would go off, and in the brief light—the last thing a couple of the attackers would ever know—Snow's men could see other forms creeping nearby toward the old foxholes. They'd blast away at them. These little VC guys weren't so frightening after all, but there sure as hell were a lot of 'em. They kept coming and coming.

The radio operator told the first sergeant that the base camp was on the radio and wanted to know if he wanted to call in the gunships. Snow told the radio operator to turn down the fucking radio. It was making too much noise. For First Sergeant Snow, it seemed to be just another relaxing evening's diversion after a full day's road construction work.

Sometime well after midnight, the VC stopped attacking. Sergeant Snow could get some of his beauty sleep.

The next morning when the patrols went out to set up fresh mines, they found twenty-four enemy bodies and no dead Americans. Snow had the bodies lined up along a taut string in a neat row so everyone could admire their night's handiwork. Then a couple of choppers arrived. It was the colonel, who wanted to know why the camp hadn't answered his radio queries the previous night.

First Sergeant Snow apologized but said he'd been very busy, as the colonel could see by the row of bodies here. First Sergeant Snow said his camp had a new policy of taking care of its own troubles and not involving others when it really wasn't necessary.

The colonel looked at First Sergeant Snow. He looked at Snow's men lined up crisply, quite a change for this unit. Then he looked at all the enemy bodies laid out along the taut string before Sergeant Snow's silent troops.

"First Sergeant?" said the colonel.

"Yes, sir?" said Sergeant Snow.

"First Sergeant, that body there, the sixth one, is slightly out of line."

"Yes, sir," said a smiling Sergeant Snow, and he motioned a man to properly realign the evidence.

It wasn't the end of the war. It was just the end of one story of the war. "We got tougher," Snow tells his listeners, "and you know what happened? Pretty soon they stopped hitting us every night. They went somewhere else for their fun 'cause I kicked the shit out of them." And then the maintenance supervisor put away his handkerchief.

Sitting there in Snow's office, in his pickup truck, or out working at Snow's ranch helping to erect a ramshackle barn with used lumber, Steve Jenkins was always fascinated by Snow's tales, especially about how the former first sergeant tamed his fear.

"Of course you're afraid," Snow would say. "You're shitting lemon juice out there." Then he would add the lesson: "The biggest fear in war is showing you are afraid, especially if you're a leader. But you can't show it. Never! I mean I'm walking around out in front of everybody with the mortars coming in, with no flak jacket, carrying nothing but a goddamned stick, hoping like hell there's no VC over in the trees yet, talking tough and telling everybody in their holes what they're going to do, swearing up my own barrage. Sure I'm scared. Only a goddamned fool isn't scared. It's how you control it that matters. You can't let it show." And he says the last sentence very, very slowly so the weighty words have time to sink through a new recruit's very thick skull.

Because Steve and Jim didn't get along all that well with other workers—some said Steve acted as if he was "two bricks shy of a full load"—the sergeant assigned them to a major renovation project at a former high school being converted to a citywide sixth grade. None of the workers wanted to fool with tearing out the old furnace and pipes, which were wrapped in dangerous asbestos. But that didn't deter the Jenkins men. They just tied scarves around their mouths and waded into the demolition work.

Snow loved them. They were inseparable. Snow had even put Steve on the payroll, part-time, since he was doing so much work. Then, when the builders came in to start fixing up one end of the school while Snow's men were tearing down the other, Jim got a second full-time job, as a night watchman for the builder. Instead of coming in at 8:00 A.M. for his maintenance job, Jim would start at 7:30. He'd take a shorter lunch hour. He'd get off at 4:00 in time to start as night watchman at 5:00. Presumably he slept sometime, but no one ever saw him. The way the builder programmed Jim's payroll, he was getting some overtime pretty near every day. But one week, after Jim's job classification had been reduced, the know-it-all machine spit out a paycheck at the old, higher rate. Immediately that morning, Jim appeared in the construction office ⟩ inform Norma Jean Parker, the secretary, of the error, and the overpayment was deducted from his next paycheck. Jim said he had to earn a lot of money so's he could go back to Minnesota and buy another farm. But he was going to earn it honest.

Jim had a reputation around the construction site for being mild-mannered and a very hard worker. Quiet but never late, never absent. He always did what he said he'd do. At times he might even point out to Arvil Gardner, the superintendent, some task that could be done better. Sometimes Jim would stop by the construction or maintenance offices to chat with the secretaries, who found him quiet and nice. He'd tell the women he had cancer, took a couple of treatments for it down in Mexico City, he said. He'd also talk about farming and Minnesota and how his wife had run off with another fellow and this big banker had taken away his land even though he'd made the payments. The women would sympathize; those problems were not uncommon in their lives. He made the Minnesota land sound awful pretty and big, like Texas only wetter. Sometimes, if she were single, Jim might sort of ask the woman out on a date. Not a bold request that could be turned down bluntly. Just a suggestion she could pick up on or not. Jim might offer to fix up her car and say they could go to a movie after or something. The woman would smile sweetly and refer offhand to her boyfriend, who might really exist.

Some days Steve would be there trying to turn the conversation to guns. When Norma Jean mentioned once that she had a .38 pistol, Steve thought that was neat. He said he wondered what kind of hole that would make in a body or piece of plywood. The conversation came to an abrupt end. After a few months, Steve left his construction job; fellow workers said he was too destructive. But Steve continued to hang around anyway, helping his father. They were rarely apart.

When Jim got in trouble one day, Steve went along. Jim had seen a teacher pull a misbehaving student from a classroom over at the junior high school and go to spank him in the hall. A stern Jim stepped in. He said that wasn't necessary, that he had never had to spank his boy. The woman, terrified, notified the principal, who notified Superintendent Lancaster, who summoned Jim immediately. Steve went along.

"What are you doing here?" asked Mr. Lancaster.

"I come to see the fireworks," replied Steve.

"Well, there aren't going to be any fireworks," said the school district chief. "Your father is about to get fired."

As they talked, Jim was not defiant. In fact, he seemed recalcitrant. Mr. Lancaster got to thinking about Jim's good work record and the hard times he knew the man had had and, of course, his incredible persistence with the tractor. So he didn't fire Jim; he just gave him a very stern warning. Jim thanked him.

The Jenkins men ate most of their meals at the Kountry Kitchen. From the three-story brick school under renovation near the railroad underpass, it was a two-minute walk up busy Austin Avenue past Lindsey's burger stand where the city cop always parked to catch school zone speeders, then past the Church of Christ and some crumbling, old storefronts, and into the brightly lit little restaurant with the jukebox and the cheerful waitresses, Jackie Foster and Mary Graves. The women took turns opening the place at 6:00 A.M., and minutes later, as if he'd been watching for them, Jim would shuffle in with that distinctive, rolling overweight gait of his. He was quiet. He'd always go to the same booth in the back. He'd sometimes have a hamburger, but he seemed to live on scrambled eggs and toast and a glass of milk. He said his stomach was always bothering him for some reason, and his teeth, too. Some days he ate all three meals there, all three of them scrambled eggs and toast with a glass of milk. Usually his kid was with him. But if he wasn't and if things were slow, Jackie would sit down at the table with Jim and they'd talk. Jim struck Jackie as a lonely man and she wanted to comfort him. He told her about his love for Minnesota. He said he was going back someday and succeed. Jackie understood that; she would've rather been somewhere else too. "But you know how it is these days," she says, "you gotta go where you can work. And that ain't usually home."

During his limited free time Jim was often with Ted Beard. Like a growing number of Americans, Beard is a former farmer. He knows how hard they work, how harsh Mother Nature can be, and what a gamble the whole business is. That's what it's become too, he knows, a business, not a way of life anymore. If you're going to be in business, you might just as well be in a reasonable one so you don't dry up over the years in frustration like he'd seen his daddy do. "I picked enough cotton to make all the clothes I'll ever wear," Beard likes to say. He got into the construction business, pulling heavy machines and his family all over the state for years. Then when his boy, Wayne, got old enough for school, Beard announced they were going to settle down in a nice little city surrounded by country where the families were strong and values decent.

He chose Brownwood and never regretted it. He worked with his gnarled outdoor hands in an appliance repair shop, eventually becoming the manager. Oh, sure, he could make more money in Dallas, but money's not everything. And he tried to teach his son that. "You know," he'd tell Wayne, "you can't just take all the time and then move on without giving something back in return—to each other, to the land, to whatever. That's what this country was made of, wasn't it? People helping people, people giving to each other, and then getting, instead of taking, taking, taking all the time."

Ted Beard liked Jim Jenkins right off. One day Jim got to talking with Ted's sister-in-law Joyce standing by a microwave oven in a convenience store while waiting for a bus. They got to dating and Joyce brought him over to Ted's house for a family picnic one weekend. Ted liked Jim's soft-spoken plainness so much that he recommended him to Jim Perry, his longtime hunting buddy who was looking for a good worker over at the school maintenance garage. That's how Jim got his job in Brownwood.

He and Joyce eventually stopped seeing each other though. She had a son, Benton, who was Steve's age. The boys hung around together. While Steve didn't try to smartmouth adults—he was always especially respectful to the ladies—he gave adults an eerie feeling. A kid that age still playing soldier. Benton had started talking more about guns and shooting after he met Steve.

Joyce blamed Steve for that, and didn't like him much. Jim got angry with Joyce; Steve was his son, his only son. "Jim thought the world and all of Steve," Beard remembers. "He'd do anything for him. Sometimes it was hard to tell who was the father, Jim was so eager to please the kid." Beard, looking at the boy and his dad, would get to thinking about his own boy, Wayne. Little boys have a way of getting headstrong as they grow older, especially around their fathers. Not all of them, of course. But Wayne was one who did. He drove like the wind, that kid. He was all for getting out of Brownwood for the big city. Wayne was up in Dallas back in January 1981 looking for work like everyone else, driving on the expressway. It was misting rain. One car passed Wayne but cut back in too soon. Wayne swerved his car. The wheel caught on the guardrail and he flipped over two or three times. The other driver didn't even stop. Wayne was twenty-five that day, about half his daddy's age. Twenty-five was as far as Wayne got.

Ted Beard worked very hard for a long while after that. Thank God for work. So he understood Jim Jenkins's preoccupation with work after the divorce. Jim never said his ex-wife's name, though once he called her "a sorry bitch." Ted didn't see it as his business to pry.

Jim seemed to fit in with the Beard family. Ted's wife Dorothy made him elk steak once, and Jim exclaimed over that. He'd never had it before, said he'd never liked hunting much and his eyes were bad too. At that dinner Dorothy said something about wanting to get extra stereo speakers for the big radio in the living room and put them all over the house. She thought that would be dandy so you wouldn't have to turn on radios in every room. The next day when Ted got home from work, his wife was all excited. Jim Jenkins had shown up that morning with his boy and they had climbed up in the attic and all over the house stringing stereo speaker wire to nearly every room. Jim had left, said he couldn't hang around for dinner, but now all Ted needed to do was buy the speakers and plug them in.

Ted was impressed with that. Jim didn't talk much, but he sure let his actions speak for him. Like the time he asked Ted to help him find a used pickup to buy. It had to be white. Then the bank said Jim would need a cosigner on the loan, and Jim asked Ted. Ted said sure, Jim's word was good. He didn't even ask how much the note was for. If Ted trusted you, Ted trusted you. Sure enough, within a few months Jim Jenkins had the loan paid off. The Jim Jenkins Ted Beard knew always did what he said he'd do.

The Steve Jenkins that Charlie Snow knew always wanted to know more about the military. He had the manuals, but he wanted more than that. At one point he prodded Snow about target practice. First, said Snow, beginning the First Sergeant's Litany so familiar to every former military recruit, you learned how to set your rifle's sights. One notch up, fire a couple, another notch, fire a couple. Check your ammunition. Check the firing range. Now sight on the target. No, no, not like that. Don't grab the trigger. What the hell are you afraid of? The rifle is pointing away from you. Now, get back down. Spread your legs. More. See, that gives you stability. Now put your elbow down. Good sturdy elbow. Now wrap your arm through the strap like this. Holds the piece tight, see? You don't wave a rifle around loose. Now, put the target on top of the sight bar. On top, not behind. Now fire once. Okay. Were you listening? I said put the target on top of the sight bar. The fucker should be standing right on it, not behind it. Now do it again. Not bad. Okay, now stand up and fire. Plant your legs like this. Put the butt up against your shoulder. Wrap your arm in the strap again. Good. Now look down the barrel— hold your fucking elbow in close, boy. You're not a bird. The elbow supports everything, makes it steady. Now, look down the barrel and squeeze off a round. Did I say jerk? I said squeeze it off. Take a little breath, hold it, now aim, and squeeze the trigger gently. Don't close your eyes. How the hell you gonna sight again if your goddamned eyes are closed? Okay, try it again. Little breath. Hold it. Aim. Squeeze. There! See? Bull's-eye! Now do it some more. What? Well, I never shot for the fucker's head. It's too small a target. Aim for the body, the big body. At least you'll wound 'em. Then if someone comes to help, you can get him, too. Okay, try it some more.

Snow noticed that Steve had some tattoos, just like the sergeant. On his right arm Steve chose drawings of a pair of crossed battle-axes, Sylvester the Cat, his first name, and the Special Forces insignia. On the left arm was his Social Security number (477-66-1558), which the military now uses for soldiers' ID tags. One payday Steve proudly walked into Snow's office with a rifle. He had bought it with his earnings. Snow recognized the M-1 carbine right off. Snow took a look at the kid's new toy; it was light, all right, a reproduction, not the real thing, but lethal enough to put a neat, fatal hole in a target.

Snow invited Steve to his ranch to try out the piece. The seventeen-year-old leftie was a very good shot and took the rifle with him just about everywhere. He'd rigged a military-type shoulder strap on the carbine just like the real thing and he'd wrap his arm in it like a real soldier to aim tightly. Steve could skip a tin can across a field, *bam, bam, bam,* just like that. And he reveled in Snow's compliments. The rifle had developed a stubborn ejection problem. It would jam now and then, despite several repair attempts. But Steve had worked out his own rapid routine to unjam the shell and start shooting again. And Snow had a job for Steve. He told him to come out to the isolated ranch anytime and shoot all the skunks, armadillos, and rabbits he could find. They dug holes that cows could step in, and they ate vegetables.

Snow isn't a big hunter himself, but he is a keen student of nature, human and otherwise. "Didja ever notice," he says, "that prey have their eyes on the side of their head and predators have their eyes on the front? Now, you look where man's eyes are."

Lots of times Snow would drive down the dirt road to his place and get out to open the gate, and he'd hear gunfire back in the trees. He'd drive in, and through the dusty windshield, off in the distance, Snow would see Steve in his camouflage clothing, sometimes wearing a full backpack and a steel helmet, running through the grass holding his rifle, throwing himself on the ground, assuming the proper prone position, squeezing off a couple of rounds, rolling over, squeezing off a couple more shots, and then jumping up to run zigzag across the field and fire, standing behind a tree. The kid was in great shape, an eager learner, a super shot. Snow felt good. He had given a healthy dose of self-pride to worse little shits than Steve and molded them into men who could be fighting men. Steve would also go out to the firing range with some of the guys from the maintenance garage, and they too were impressed.

Late in the spring of 1983 Steve Jenkins disappeared as quietly as he had come. His father said he'd gone back to Minnesota.

Jim kept on working both jobs, night and day. In midsummer he walked into Snow's office and told Snow he was overtired. He said he was going to quit his maintenance job and just work as a watchman for the construction company. Snow told him to think about it carefully. That guard job was temporary, he said. They have one at every work site. As soon as the work was done there, Jim would be laid off. Jim said he already had thought about it. He wanted to quit. Snow shrugged his shoulders, and Jim left.

A few days later Jim came back. The work had finished at the construction site, he said, and he'd been laid off his watchman's job. Could he have his old maintenance job back?

"I told you what would happen," said Snow. "Would you hire yourself back after that?"

Jim thought about that for a moment. "I don't believe so," he said.

Snow was touched, although, of course, he said nothing about his feelings. He had already hired a new man. For a second, Snow thought of hiring Jim back on the spot, just for the honest answer. But before he could open his mouth, Jim cut in.

"It don't matter anyhow," said Jim. "I'm going back to Minnesota to farm again." He pulled from his back jeans pocket one of those big zippered wallets connected to his belt by a chain. The wallet was bulging at the seams. Jim said some men dream of having a few hundred dollars in their pocket. Now for once, he said, he had several thousand dollars in his pocket. He was gonna go get him another dairy farm. This time, he said, it was gonna work for sure.

Soon after, Jim appeared as usual at the Kountry Kitchen to eat. He told Jackie and Mary that he was leaving Brownwood to go back to farming in Minnesota. They were very pleased for him. He said he was sorry that he had never left them a tip after all those meals; he'd been so short of money saving every penny for the farm. Now, Jim said, he wanted to make it up to them. They should pick a time and place, and he would take them both out to dinner. They chose Saturday night and the China Inn, which is just down Austin Avenue in the Village Shopping Center next to the Perfection Salon and Clint's Red Lantern bar. The restaurant featured chow mein for $4.75 and an entire seafood dinner for $8.60. It has private booths with vinyl upholstery, a tiled floor like the fancy places, and real Chinese characters carved in stone out front. Saturday night right on time, as the women closed the Kountry Kitchen and removed their aprons in an excited hurry, Jim drove up in his pickup truck. The white truck was washed clean. Jim was wearing a new Hawaiian shirt.

The next morning bright and early Jim set out on Route 377 to Dallas and on to Interstate 35 for the long trip north, back home to his future.

At about that same time, in 1983, Rudy Blythe was on the phone from Ruthton to his wife in their Dallas home near the interstate. There were some serious problems at the bank, he said. He wouldn't be able to sell it for a while. He hoped Susan understood. Could she come home?

First of all, thought Susan, Ruthton is not my home.

But then, neither is Dallas. Rudy had been spending more and more time back in Minnesota, actually full-time for several months. As a result, Susan had been leading the celibate life of a single parent in Texas. She'd even had to get a job there, helping to manage a savings and loan branch. Rudy said they needed her eighteen-hundred-dollar-a-month salary to help cover the Texas mortgage and living expenses. Rudy was pouring everything else back into payments on their bank just to cover the interest.

Susan had noticed that Rudy was talking louder and quicker lately. When he did snatch a few days to visit her and Rolph in Dallas, he tossed and turned in bed all night. So Susan didn't argue about returning to Ruthton. Maybe it wouldn't be so bad this time, she thought.

5

THE HEAVENS' ARTILLERY

Rudy was worried. He was in trouble, deep trouble, and he knew it now. His bank might be failing; certainly its capital base was eroding. He was failing. The parts wouldn't go together. And there was no one around to help. If other rural banks were also quietly feeling the same frightening tightening in their stomachs, that didn't make it any easier for Rudy. His dream was fading. Or was it a nightmare?

Rudy had long known there were problem loans at the Buffalo Ridge State Bank. Every bank has them. Every bank expects them. When he worked in Dallas and visited his Ruthton bank monthly or chatted on the phone with Jerry Ihnen, Rudy was detached from the problems. During that year he knew about some of them, but they didn't seem so pervasive from a distance. Jerry was hired to run the bank. And he did—except, of course, for those increasingly frequent times when a nervous Rudy overruled him from afar. In the fall of 1982, Jerry made an announcement from afar: he was leaving to start his own insurance agency.

That bombshell was delivered to Susan by phone with two weeks' notice. Rudy was angry, to say the least; he would never have another nice thing to say about Jerry. More important, whenever Rudy was angry, he was frightened, and this time Rudy was very frightened. While Jerry ran the bank's day-to-day operations, Rudy had gone to the new job in Texas to earn enough money to continue carrying the bank loans. Now, not only was Rudy losing the man who ran the bank, making it possible for him to stay in Texas, but this man was taking with him intimate, detailed knowledge about Rudy's insurance agency, whose steady flow of commissions provided the financial foundation to pay the bank loans.

As a result, Rudy quit his Texas job in October 1982. The Dallas bank said it would hold his position open for six months while Rudy tried to sell the Ruthton operation. Susan would stay in Dallas with Rolph and her volunteer work and, very soon, her full-time job at the savings and loan. She expected the Minnesota bank to sell any day.

Now, instead of commuting from Texas to Minnesota to check on the bank each month, Rudy was commuting from Minnesota to Texas each month, trying to balance a financially troubled bank and a financially troubled personal life with an expensive house and a family in Texas. Then the Texas bank wanted its bridge loan paid back. It was one thing to give an employee a temporary loan to buy a local house. It was quite another to leave that interest-free money outstanding to a former employee who was having trouble freeing himself from his old business in a rusting region with no future. Once again, Rudy had to turn to his father for financial help. Rudy had only one option.

Going back to work for a while didn't bother Susan. That was part of the bargain in a modern marriage. Hadn't Susan's parents back in New York brought her up to be independent and resourceful, maybe not an obedient wife in the traditional sense but certainly an understanding partner? Rudy had gone to Texas for her. And why would Rudy feel threatened because his wife was earning so much money to support the family? Anyway, when Rudy visited Texas during the coming months, it was like a vacation in many ways, and it was good to be needed.

As soon as Jerry had given his notice, Susan began looking for a replacement. It was hard to do from Texas, but Rudy was busy running the bank by himself. Susan placed ads in midwestern newspapers and wrote letters to corporate recruiters. She interviewed candidates by phone, and then checked with their last employer. She drew up a list of candidates for Rudy. There weren't all that many, to tell the truth. In fact, the only serious possibility was a fellow named Deems Adair Thulin, not Too-Lynn but Too-Lean—call me Toby.

He was a Minnesota boy, thirty-seven years old, a Vietnam veteran the application said. He lived with his wife, Lynnette, and their three little girls—Linda Marie, Deanna Lynne, and Kara Anne. He'd worked for over ten years as a bank insurance salesman and loan officer in a couple of little Minnesota cities. Now he was in northwest Iowa, unemployed. It seemed strange for a guy like that, so Susan called his last employer, the Sibley State Bank.

The man at Sibley seemed reserved in his endorsement. Toby was a decent family man, he said, very active in the community. He'd worked for them only about a year. There had been a mutual parting of the ways.

What was the problem? Was he an embezzler, a homosexual?

Oh, no, said the bank officer, nothing like that. It was mainly a personality conflict.

Big deal, thought Susan, who'd had ample experience with smalltown personality conflicts. So she recommended that Rudy interview Thulin. They were in a hurry for a replacement, and Rudy would be there full-time now, at least until they dumped the bank. As long as it was his, Rudy was not straying far from the bank; hired hands, he'd learned, don't care the way owners do.

Rudy liked Toby right off. They were both tall, over six feet. Toby was also from outside Ruthton. An ordinary guy. He'd had a lot of loan work. A fellow veteran. He had only one year of college, so he was no intellectual threat to Rudy, who was no intellectual. Like Rudy, Toby liked sports, though he favored softball with the firemen over golf with the bankers. He liked the outdoors, hunting, small towns, and his children. Rudy hired Toby immediately, and Toby began work in November 1982.

The new man's first assignment was to help Rudy wade through all the bank's loans. There were hidden problems and Rudy knew it. For one thing, he had a substantial number of borrowers who were behind in their payments, men who had guessed wrong on economic conditions and interest rates. Rudy knew this because he had guessed wrong himself; the institution that loaned him the money to buy the bank had flagged Rudy's loan as a problem and was watching him more closely. Now Rudy had to do the same down the line.

If Rudy's borrowers weren't behind in payments now, they could be soon. Maybe he'd better move to get his money out of these guys before they went under and took his investment with them. Maybe they didn't know interest rates would stay up so high so long. Rudy thought maybe they had tricked him when he was so eager to loan locally. Rudy hated being tricked. He had come into town with the best of intentions to help, and now it seemed a fair number of these good ole boys in dirty baseball hats were opting to pay other debts before the bank. If they had anything left at the end of the month, they might put a little down toward the bank loan.

Well, that just wasn't good enough. Rudy was going to have to get tough. He and Toby would methodically go through the loan files. They would visit these guys on the farms now and then, and ostentatiously inventory the collateral. There were going to be no more cute little Jenkins tricks, selling the stuff themselves and then raising their hands helplessly as they walked into the protection of bankruptcy and left the bank holding an empty mortgage bag. Rudy was still stuck with that Jenkins place.

Rudy could see it happening all around him. Farmers going under or walking away. The sales notices went up on every bulletin board around. For every few farmers gone, another business in some town would collapse too. The bankers would quietly inquire among themselves as to who held the loans for that guy, and when they learned what bank it was, they'd shake their heads, professing sorrow but really feeling relief. Thank God, it wasn't them this time.

As more people went under financially, those remaining became more frightened. The more risky financial life became, the more information bankers wanted. The more questions they asked, the more cautious their clients grew. The more surprising the farm and bank collapses were, the more suspicious both sides got. The more trouble those in debt reaped, the less desirous those without debt were of buying. The more farm machinery and farmland and houses went on the market, the less each was worth.

Rudy and Toby sat down and figured out what that tractor and this land and that corn dryer were really worth if sold today. They discovered, as all the local bankers were discovering, that the declining value of the collateral often was no longer equal to the loan. Something had to give, and this time it wasn't going to be Rudy's dream.

Week after week this painful adjustment process went on. When Dick Ness stopped by the bank to chat with Rudy one day, he thought he'd ask for a little loan. He wanted to trade in some old farm machinery for new stuff, take advantage of the discounts the manufacturers were offering to spur stagnant sales. Ness knew he'd have to pay a higher interest rate, but they'd probably be going down soon. No, said Rudy, he didn't think they would. He was very pessimistic. The problems were here to stay, he said. He thought a lot of people would fall by the wayside in that time. That disturbed Ness. Jolly Rudy down in the dumps and talking gloom. Maybe this financial situation was more serious than he thought.

Rudy was harsher with other applicants. There were new stories of broken pencils and some swearing and yelling about stupid ideas. Sometimes it seemed as if Rudy didn't even take time to think about a proposal. Just a quick, abrupt decision. No! Which is spelled N-O. Period. Now get out of here.

Sometimes, to convince a recalcitrant borrower, Rudy and Toby would resort to the good-guy-bad-guy routine. They called it the white-hat-black-hat method. It is a procedure as old as the world's police forces. And they'd take turns wearing each hat. One banker would go into a meeting with a borrower like Attila the Hun, talking higher interest rates, no more loans, courts, foreclosures, seizures, attorneys. The other would try to calm his partner down, noting the client's intelligence, goodwill, presumed intention to repay, how tight times were all over. But the black-hat banker would remain unconvinced. Why should he believe this guy now? He was months behind in payments—what was it now, four months behind? —and he was always promising to pay something next month, next month, and then he didn't. Well, next month had come, you know? Things were getting ridiculous around here! He'd been understanding long enough. This wasn't a goddamned charity operation. The bank had obligations too. No one had forced the farmer to take that loan.

Then the black-hat-man-of-the-day excused himself from the room for a moment. The white hat would raise his eyebrows and apologize for his friend but say he hoped the borrower realized how serious everything had become, and maybe, if this client could come through with a hefty payment right quick, you know, get everything back on track and schedule, then he could calm his partner down and everything would be okay and back to normal.

Faced with being closed out in ten minutes or getting to keep his existing demand note as long as he brought the payments back up to date, which was no more than he had agreed to in the first place, the borrower's decision seemed a simple one. He could probably scrounge up some money somewhere and make a big payment, maybe tomorrow, if that would be soon enough. Or maybe, if he was a little outraged by this angry behavior, the borrower would say that he didn't have to put up with this, that he'd been a regular customer for three years, and he might just take his business to another bank. You know, get a loan from a Pipestone bank, and pay off Buffalo Ridge, and be done with this kind of foolishness. Who needs this kind of treatment, after all?

The white-hat guy would say, gee, they sure would hate to lose his business, but it was a free country. And, yes, he had heard that the Pipestone bank was making good loans and in fact he happened to have their phone number right here. But either way, you know, things had gotten serious and something would have to be done about this payments. Nothing personal, you understand. Just business.

At other times, big Rudy, wearing a coat and tie and carrying his worksheets, would go out to visit a farmer, who was wearing a plaid shirt and dirty coveralls. Rudy would sit down at the kitchen table with the farmer, who knew this banker from somewhere in the East hadn't stopped off by chance. The missus would serve some coffee and homemade cookies and then disappear into another room, keeping the children and the sniffing dog at bay, but keeping the TV volume low and a sharp ear out.

The men would talk about the weather at first, whatever it was, too cold, too hot, too wet, too dry. There would be an awkward pause, and then Rudy would clear his throat and get down to it. He'd gone over all the figures for him, the loan, the interest, maybe some missed payments. He'd listed all the assets, the farm, the machinery, the stored crops. He'd projected, conservatively of course, what the place would produce in crops and income that year. And, well, frankly, things didn't look too good, not good at all, as the farmer could see on this cash flow sheet right here.

Of course, things weren't too good anywhere. Crop prices were down. This damned, excuse me, interest rate was staying way up there. Nothing anybody could do about that for the moment. It was hurting the banker just as much as anyone else. Rudy had payments to meet too, you know, but the fact of the matter was there was just more outgo than income on this farm, plain and simple. Something had to be done. And before things got worse.

Now, Rudy explained, the farmer could decrease the outgo or increase the income, or both somehow. One way would be to cut down the debt. Sell off some things, machinery or maybe a little land, and put the money toward the debt, get it down to a more manageable level where he could handle the payments at today's rates. Now, prices aren't too good these days. But the way they're going, they aren't going to get any better in the future. It would be hard, Rudy knew, to sell off some of this land that had been in the family for several generations, but farming was a business now. No room for sentiment.

The alternative, of course, would be to get more income. There didn't seem to be any way to do that on the farm; crop prices sure weren't going up, not since that grain sales embargo after those Russians invaded Afghanistan. But there was one way to increase income: the farmer could send his wife out to work.

Well, now wait a minute. It's not that bad. Wives are working all over the country. In fact the banker's wife is working right now. Down in Texas. And the extra income sure comes in mighty handy, pays for a lot of things. The farmwife's wages might just make the difference on the loan payments—let them keep the farm intact, get their heads back above water, at least until interest rates come down. Think about it. Something has to be done. These bad loans just can't be allowed to accumulate. Else there won't be any money coming into the bank to make new loans. Nothing personal, you understand. Just business. Even bankers have to make their own loan payments.

Toby was moving through the bank's loan lists alphabetically, examining each, updating values, checking for troubled ratios of debt to equity. He didn't know the personalities involved and didn't like the arm-twisting part of banking at all. Have you gotten to so-and-so yet? Rudy would demand. Rudy wanted him to work faster, to share his sense of urgency. No, Toby would say, I'm just up to *H*. Rudy wanted things in order, and quickly, if not to sell the bank, then for his own peace of mind. So he'd have Toby jump ahead in the list to suspected trouble spots.

Toby, a former air force sergeant who used to talk often of his dislike for high-and-mighty officers, would tell his friends that he hated playing the role of enforcer. It upset his stomach, even when the borrower paid back some of his loan. Toby was smoking more. Maybe he'd say it reminded him of his days as a prisoner of war in Vietnam undergoing intensive interrogations by Vietcong commanders. At least the bank didn't hang these borrowers up by the heels and stick bamboo splinters under their fingernails and into their body openings.

Whenever Toby knew he was going out to visit a muddy farm, he'd wear his combat boots. He joked about that but told Lynnette several times he was worried about the mood in the countryside. It was turning ugly, he said. Toby said he'd never seen this kind of militance among farmers. He said he was afraid. But Toby had an array of lingering debts himself, bad enough for anyone entering a new job but doubly embarrassing for a banker, if anyone knew. So Toby didn't have much choice about enforcing collections for the Buffalo Ridge State Bank. He needed the salary. He hadn't had any work to speak of for more than a year, a time when his impoverished family sometimes discovered small envelopes of money in their coat pockets after leaving the church cloakroom.

It had been a very trying period for Toby and Lynnette, who had worked their way up from being newlyweds on a Georgia air force base to living in a house trailer to building their own home, with the dream of someday constructing a log cabin together in the northern woods on a lake. Lynnette had seen the lack of work and the financial hardship corrode her husband's confidence and their relationship. He was spending more time away from home; he always seemed to have something to do with the volunteer fire department in Sibley or his nearby National Guard unit. He'd mumbled about not even being able to support his family anymore.

And he started in again on all that Vietnam stuff, the nightmares about women and children coming at him with knives. One midnight Toby was yelling and thrashing so much in bed that Lynnette sat up to wake him. Just then his fist came flying through the darkness and slammed into the pillow right where her head had been. That frightened Lynnette.

As a young woman, a bank teller from a very religious family who lived in a series of little midwestern towns, she had set her sights on a tall, handsome young man with curly brown hair whom she'd met at summer camp. For young Lynnette, the Vietnam War had been a place to send cookies to. She wrote letters to Toby every day. He replied, religiously by his standards, every other month. He told her that the fighting in the Special Forces had been so fierce and so isolated that he only got her letters in large bunches upon returning from the long patrols. When she finally wrote him a "Dear Toby" letter and said to forget ever seeing her again because he couldn't even be bothered writing, he telephoned her from the faraway hospital where he said he was recuperating. He proposed to her. Lynnette had said yes, of course she would marry him. She felt very badly about her angry letter.

Toby had also spent most of his life in a midwestern small town. He was a child of the post-World War II baby boom. His father was in the navy on a fleet oiler when the American flag went up over Iwo Jima. Toby was one of seven children born to Larry and Camilla Thulin of Ada, Minnesota. Five of their children survived past infancy; one was retarded.

Toby's father was a hardworking contractor who called his son Toby because it was his own navy nickname. Toby's mother was a hardworking music teacher who chose the name Deems for her son because of her admiration for Deems Taylor, the American writer, painter, and operatic composer. The name Toby stuck, however, and he emerged as his father's favorite child for the love of the outdoors and hunting that they shared. As a boy, Toby got into more than his share of mischief, climbing on tall scaffolds, breaking a few high school hearts, and downing a beer or two on the way to a C average in school. According to his mother, Toby's childhood was not the most peaceful. There were the deaths of two siblings, one only two years old and a favorite of Toby's. And she said her marriage was disintegrating bitterly, although in those days the only socially acceptable thing to do was persevere until the children were grown, divorces being for Californians and other rich folk. So in the Thulin household there was lots of yelling, and the children had to choose sides.

Toby's mother, who was still teaching as she approached her seventies, believes in studying music as one of life's essential disciplines. So Toby studied music, practicing one full hour every single day. He played piano for a few years and then violin with a depth of feeling and musical sensitivity that belied his youth. He worked his way up in the high school concert orchestra until he was to become concertmaster. But that year the high school had a German exchange student who happened to play the violin. Out of politeness, he was given the lead chair. Out of disappointment, Toby Thulin gave up music. He said he wanted to become a fireman.

He still tried in track and swimming. He would practice and practice, running extra sprints, swimming extra laps. Then when the big race came, Toby would finish third or fourth. Never first, never the star, no matter how hard he tried.

At his parents' request, Toby went to Nazarene College for one year, before dropping out and joining the air force. Finally, it seemed, Toby found success in the service. His mother fondly remembers his stories about the dangerous foot patrols in the jungles of Southeast Asia and the heat and then spending weeks teaching English to village children. Toby's stories were even better than his father's war stories. The senior Thulin remembers Toby's tales about being a gunner in the door of one of those C-130's, the fixed-wing planes they called Puff the Magic Dragon because their Gatling guns fired so fast, they seemed to be spitting white-hot lead. What Toby's father would have given for one of those guns when the kamikaze planes came after his fuel ship in 1945!

The father was very proud of Toby's exploits. Toby was a star in his father's eyes. The old man told him so the day the kid telephoned all the way from Bangkok over there in Vietnam. He'd just come in from a jungle patrol, he said, and he wanted to hear his father's voice. Toby's mother didn't get a phone call, but her son did bring her a radio. It was from the PX in Saigon, he said. Thank God, Toby had come through all that fighting unscathed. Her stomach began to feel sick when she saw that plane take off to the west with Toby, and she didn't feel better until a year later when he strolled into the living room all tall and blue in his uniform.

They tried to convince Toby and Lynnette to postpone their marriage until he had a steady job. But you know kids these days. They got married right away anyway. After his son's discharge from the air force in 1969, Toby's father gave him a job with his company. He hoped the boy would work his way up within the firm and take it over someday. Toby liked the rough company of working men. He spent a lot of time with various groups—the National Guard, volunteer fire departments, emergency rescue squads, deputy sheriffs' patrols, scuba diving club, fishing friends, and hunting buddies.

Toby loved hunting—birds, deer, whatever was in season. He was a very good shot. You had to be in Vietnam, he'd say; you didn't get a second chance. He didn't talk with these men too much about the war. He implied a lot, which was proper; he couldn't be seen to be bragging too much, coming on too strong for the friends who hadn't braved combat but now found themselves united in the chilled woods in their affinity for guns.

Every fall hundreds of thousands of midwesterners go hunting and killing, much of it legal. They schedule their vacations and shifts around it. They buy their ammunition and beer at the gas station. They load their rifles and shotguns. Calling themselves sportsmen, they walk off into the woods to deal death to prey. Many times they bring along their young sons, who may well have gotten their first guns not long after learning the awful truth about Santa Claus. For them, these trips are a rite of initiation into manhood, or at least part of the process, along with drinking beer, driving cars, and making love, all begun with fumbling steps and, later, tall-tale telling.

The hunting father feels good leading the way, teaching what he knows, being looked up to. The young man feels older to be included and learning among friends, who may smile silently at a mistake but will never laugh. The men were ten too once. The women seem to opt out of these excursions, unless it is important for one to feel included by a special man. Fishing may be okay, the actual hooking of the cold-blooded victim coming far below out of sight. But carrying a gun that issues that brief bark of death is something left to stronger stomachs or different minds.

The males will sit in the boat or the blind or steal through the woods trying to be quiet and to forget the cold, soaking up the camaraderie as well as the water. The youngster with his small rifle may be able to blow the head off a crow or squirrel just for practice. The men will cheer the fatal achievement with good words and kind pats and helpful tips on better aiming for the next target. Men especially seem more comfortable around death; it is an end with no threat. A birth can augur anything a fearful eye wants to see.

It is not the issuance of death that appeals to most hunters. Some, privately, will even admit to a sadness. What appeals is the tribal bond of friends and the skill or luck of the hunt, outfoxing the fox. There is also a palpable sense of power, to stand here and by merely bending a finger, cause something sudden to happen over there. For those who are not very important elsewhere, that can be very important. Killing on a hunt is one of the few times a man can feel in complete control, undefied, unvexed by modern complexities, for once determining absolutely the future without regard, for an instant anyway, for any annoying complications. Even if it is to destroy a future. It is not always just exploding gases and pieces of copper and lead that hurtle down those lonely, steel barrels. The bullets can be propelled too by hostility and frustration.

Few hunting victims ever know what hit them. There may be a split-second of silent fear, a moment of foreboding when an unusual noise in a grove of trees raises instant suspicions, too late. Some hunted creatures freeze then, seeking to hide in immobility, their motionlessness belying the rapid pounding within. Beast or man, those are the easiest targets. Sitting ducks. Other victims try to flee, dashing in panicked zigzags there and there and anywhere but here. But bullets are quicker. And a good eye peering behind the sight in the bushes can anticipate where the vulnerable hunted will run next out in the open. A bend of the finger here and something sudden happens over there.

Nothing can quite prepare anyone, even those intimately familiar with killing and its messy aftermath, for the loud pop of death by gun. One moment there is a living creature, ears perked, eyes alert, heart pumping, mind thinking.

Then, *bam!*

Faster than the sharpest sharpshooter's eye there is death. A red meaty hole where once there was skin. In a flash, the hunted thuds down, limp and flaccid on the cold, wet ground, eyes staring blankly out, teeth showing slightly, blood and other fluids seeping slowly in waning spurts, even as the body twitches a moment.

One second, life, hope, and promise. The next second, death—in an empty shell with an awesome and absolute forever finality larger even than the broad prairie skies looking down from above.

Toby went out hunting with the guys so much that Lynnette began to complain. He was gone so much that he once missed a close friend's wedding and even little Kara's appearance in the Miss Osceola County pageant, for "required" National Guard days that Lynnette later learned were really optional.

The Thulins' stay in Sibley, Iowa, started out well. It is a typical northwest Iowa town of three thousand dominated by a blue water tower, a yellow tornado siren, and friendly pedestrians who greet most strangers with a "Howdy." "Pos-Sibley the greatest town in Iowa," says the sign out on Highway 60. Most residents have lived there for so long that they hardly glance up at the spectacular pastel prairie sunsets of reds, oranges, blues, yellows, and grays with colored clouds strung from the western horizon as if blown back by the dying day's cosmic winds.

The modern brick and glass structure of the Sibley State Bank was one of the newest structures in town, of course. The bank originally hired Toby to sell insurance. Toby spent weeks organizing a massive system of files. The bank even bought new cabinets to handle them all. Then Toby began disappearing on daylong sales trips that didn't seem to produce much business. He also sought some days off for counseling at a veterans hospital in a nearby city. He said he'd been exposed to Agent Orange and needed help, and his boss said, certainly, do whatever you need to do to get straightened around, although another bank officer swore he saw Toby going into a local store when he was supposed to be at the counselor's.

After barely two months, Lynnette, who had impressed bank officers in the initial interview, was seen as something of a complainer. Toby, too, was less polished and more blunt than anyone had anticipated. Some coworkers thought perhaps he was being pushed to be something that he wasn't, namely a bank executive. At one employee coffee break Toby began describing his favorite position for sexual intercourse, which stopped conversation. Then there was the question of the checking account. Toby didn't open a free employee's account at Sibley State. He went two blocks away to pay a fee at First National.

Toby did knock off all the drinking. He just stopped one day, and all he'd ever have again was an occasional beer. But he couldn't stop the smoking, which bothered Lynnette no end since her father was slowly shriveling with lung cancer, and breathing difficulties did seem to run in Toby's family. There were disagreements over money. There always seemed to be enough for scuba diving or a hunting trip or a new .45 pistol like the one officers carried, but Toby complained of tight finances for other things. Lynnette's idea of essentials ran more to nice, brand-name clothing for the girls and spiffy swim suits for each of them each year. One time in Iowa, Lynnette recalls borrowing several hundred dollars from her father to pay some accumulated bills and giving the checks to Toby to mail. Two weeks later the Thulins got all kinds of complaints from the merchants. Where was the money? Lynnette checked Toby's coat and there were the envelopes, unmailed. Toby said he didn't remember ever talking about the bills. And the money was gone from the checking account.

Toby talked for a while about rejoining the military, getting back with the guys and the regular paychecks, the structured life and travel. But Lynnette, who had moved around the region with her father the truckdriver, said the last thing she would ever do, other than give up going to church several times a week, was pack up the girls and move all over the country again every few months.

Then, of course, there was Vietnam. Vietnam. Vietnam. Vietnam. Lynnette was sick of that place, of hearing how horrible it had been over there. Okay, fine, it was horrible. It was more than a decade ago though—and Toby had volunteered. How long was he going to cling to those nightmares? This was the Midwest in the 1980s, and things weren't going too well right here. Either forget this Vietnam business or get some help. Whatever you need to do, just do it. But please spare her all the moaning. It has no doubt been tough on a lot of other people, and you don't hear them moaning.

Toby did stop talking about it. He also stopped talking about practically everything else. The four females in his family began to see the man in their lives as a walking time bomb ready to explode verbally at unpredictable times. Even at a friend's house, Lynnette and Toby would bicker over everything and anything. Their friends would fall silent and take a sudden, intense interest in clearing the dishes, getting out of the room, and vowing in kitchen whispers never to have the two of them over together again.

These troubles didn't show when Toby first arrived in Ruthton that November, although one woman's antenna went up when she saw Toby, the family man, working alone at the bank late on New Year's Eve. Even the bank's worried owner wasn't in that night.

In Ruthton, the Thulins began living in Rudy Blythe's big bank house, the one with the yard cluttered by repossessed farm machinery awaiting resale. With Susan and Rolph in Texas, Rudy needed only a bedroom at one end, and the Thulins had the rest of the place. It was a chance for them to start over financially. Rudy and Toby got along well, Rudy had his own tie-and-coat social circle, and Saturday dinner out, followed by a little late reading about MacArthur or Churchill. Toby was a frayed-jeans-and-sweatshirt guy, puttering around the garage or wandering down to the ball park to join a game, have a cold one, shower, and catch a TV movie or thumb through the latest *Soldier of Fortune* magazine. And anyway, he said he went back to Iowa one or two weekends every month for National Guard duty, as if it were the only Guard unit in the country.

To his small group of local friends, Rudy seemed very happy to be back in the country. Rudy found himself drawn to Lee Bush, the bank's rural attorney. He'd often eat at their old place with the canoe and the pond outside Tyler where Lee, Lee's father, and Lee's grandfather had grown up and learned about prairie life and death and the power of the weather. Before dinner, Rudy would paddle around the pond, dipping the oar and his hand in the old, cool water. He decided then to take Rolph on a two-week canoe trip in Minnesota's northern wilderness that summer. So one day when Lynnette walked in the house, there was a big dome tent set up in the living room. Rudy was inside; he said he had to practice this stuff so he wouldn't look dumb in front of his kid on the camping trip. He also tried to frighten the women at the bank, walking out of his office wearing one of those big beekeeper-type hats with the dark netting over his face. He'd heard that northern Minnesota had mosquitoes so big they showed up on radar screens.

Rudy also decided that, unfortunately, he had to sink some money into that Jenkins place. It hadn't sold, although Toby and Lynnette had considered buying it. But after seeing the holes Jim Jenkins bashed in the wall and floor while tearing out the plumbing, Lynnette said maybe someplace in town would be better after all. She also remembered Rudy telling her that one afternoon long after the repossession he saw Jenkins walking his old property, carrying a shotgun. Good thing Toby was a combat veteran.

Rudy liked organization and feeling prepared. On one of his occasional checks at the Jenkins place, he had found the roof leaking. So he hired Lyle Landgren and Dick Austin to fix it up in their spare time. They both commented on what a creepy place it had become, empty and without life for so long. Not long after that, Lyle heard an emergency call on his police radio, a fatal accident over on the highway to Marshall. That area wasn't his responsibility, but the victim was his friend. Dick Austin's time had come suddenly. Some guy in another car, maybe trying to save a few bucks on repairs, had dozed off from carbon monoxide fumes leaking from his exhaust system. The sleepy guy had crossed the center line at a pretty good clip and had scraped off one side of Dick's car, the driver's side. Dick's wife and boy were going to be okay. The story didn't make the city papers. It didn't seem worth mentioning.

One story that did make the papers, however, was the protest by a couple of hundred farmers at the sheriff's sale of one foreclosed farm up around Ivanhoe. By law the sale had to be by public auction. The farmers gathered outside the county courthouse to jeer and shout and wave signs to disrupt the sale. They failed, of course. Abe Thompson, Lincoln County's sheriff, would not be kept from his duty. Down in Illinois that spring a bunch of farmers got together with some auto worker union members and broke up a forced farm sale by drowning out the auctioneer and forcing him hard up against a combine. Out in Colorado a similar sale erupted into violence with tear gas and arrests. In Iowa and Minnesota some rural farmers and urban union members began trading appearances at evening meetings to voice their grievances about interest rates and the economy. The gripes turned out to be similar, surprisingly similar to many, and they agreed to work together on a number of issues. Up in North Dakota a farmer had gone off his nut, shot and killed two federal marshalls. A bunch of wheat farmers had filled two rail cars with grain, shipped them to Minneapolis for milling, and then had the flour distributed free to unemployed auto workers in Detroit, an action that struck some students of history as strange since farmers and union workers weren't supposed to get along.

"Susan," said Rudy on the phone, "you don't realize how serious things are." Susan, in Texas, was trying to cheer up her husband in Minnesota, but nothing worked. The bad loans were accumulating. She could tell by that tone in his voice that he was becoming more worried every day. That frightened Susan. So much so that she didn't even make a fuss when Rudy revealed that he wouldn't, or couldn't, sell the bank for some time. So much so that she readily agreed when her husband said she should move back to Ruthton, a place she hated and thoroughly distrusted.

The Blythes spent much of that summer of 1983 organizing the move, listing the house, packing, and hauling possessions back north after Susan had quit her job in Texas. Rudy continued to tighten operations at the bank; Susan helped wherever possible, even doing some cleaning. In fact, Susan cleaned the bank so regularly, a friend joked that her title should be Vice-President and Janitress. Rolph and Rudy took their wilderness canoe trip together. It was great fun, according to Rolph, a long rainy time, according to Rudy, and just fine with Susan, who preferred her wilderness jaunts equipped with comfortable cabins, hot showers, and flush toilets.

With Susan returning to Ruthton, the Thulins had to leave the Blythe house. They did, but not together. Toby had Lee Bush serve Lynnette with separation papers in June. On August 8, she and the girls left for Fargo, North Dakota, with their belongings in a rented truck that had more gears than Lynnette knew existed. Lynnette was going back to finish college, bitter but determined. Toby moved into another bank-owned Ruthton house, another repossession, in fact, from a young couple who also had separated.

Slowly word got around town. There was a new woman in Toby's life, Karen Rider, a divorcée with two children, Missy, twelve, and Casey, eleven. She had appeared in town after the Thulins' separation, moving up to the Ruthton area from some little town

in Iowa, Sibley or something like that. She rented a
house in Lynd and got a job working with handicapped youngsters in Tyler, both just up the road from
Ruthton. Karen, a redhead who wore large eyeglasses,
was forty-two when they moved to Minnesota, five
years older than Toby, and tired of too many lonely
years as a single parent. She had met Toby several
years before during a visit with a female friend to a
National Guard camp. It turned out that Toby and
Karen lived only a block and a half apart in Sibley.
They would talk often there; that's all there ever was
between them until the separation, Karen maintains.
Each was very lonely. Each sensed the other caring.
Karen said she liked outdoor activities—camping, hunting, fishing. They talked wistfully of someday building
a log cabin together in the northern woods. Toby liked
her children, who needed a man in their lives. Even
before Toby's separation he was attending some school
events, track meets, and Casey's football games. Karen
had been married sixteen years to a handicapped professor, but Toby was the first man in her life to send
her roses. He attached an anonymous card that said
simply, "You're Neat!"

Karen also was eager to listen by the hour to Toby's
stories of Vietnam. He told her about life there in the
army. He had started out in the air force, Toby explained, and then had been specially recruited into the
commandos, the Special Forces. Their training included
spending two nights at sea without a boat, he said. He
was a point man, too, the lead scout on the hairy
jungle reconnaissance patrols in enemy territory. Toby
described being captured twice by the Vietcong, about
being tortured sexually by female guerrilla fighters,
and about killing one of them, a guard, to make his
escape one night from a bamboo cage. Karen was very
sympathetic to these horrors, and she would watch
war movies with him and talk about them afterward.
Toby said Karen was so different from his wife. Toby
told that to his father, who had never liked Lynnette.
Larry Thulin always felt Toby could have done better
for a wife. And he missed their frequent times together.

*   *   *

In late August around the Ruthtons of the Midwest, the pace of life picks up noticeably. The sultry heat and humidity still hang everywhere, except in air-conditioned sanctuaries like the cafe, tractor cabs, and some bedrooms, but vacations are ending. Thoughts turn to school starting in a few days, and soon after, the beginning of the harvest, with all the pressures and obligations of those full days. As August was ending, several hundred local residents gathered not far from Ruthton for a late-summer public picnic with lots of food and cold drinks, some games, and much talking. It is a good time to see people and be seen by them. And so Mike Cable, the Lincoln County prosecutor, stopped by. He chatted up a lot of folks, ate some hot dogs, and noticed that Jim Jenkins was back in town. He'd been off in Texas or somewhere for a long time, Mike knew. Jenkins had his son Steve with him. Weird kid, shaved head, camouflage clothes, and a machete strapped to his leg. Word was, those losers were trying farming again, down near Hardwick. Better not let Rudy Blythe hear about it.

In early September Rudy had gone to the doctor complaining of occasional memory loss and dizziness. Maybe it was all the strain and long hours he'd been putting in. Maybe it was the excitement of the family about to be back together again. Maybe it was his weight, which had crept up to around 260 pounds. The forty-two-year-old banker was determined to go on a diet again. The doctor could find nothing wrong, but he suggested a specialist if Rudy was still concerned. He was. Some nights he had trouble sleeping; the phone would ring at all hours. It was some man—he didn't sound drunk—telling Rudy to make his dog shut up else he'd shoot it. Rudy would reply, accurately, that his dog was right by the bed quietly asleep, which is what all normal beings should be doing at that hour.

On September 21 Susan set out by car from Ruthton for Texas with her friend Sharon Fadness. The Dallas house had not yet been sold. Autumn's cold and storms were fast approaching Ruthton. Susan would check on the Texas house, perhaps lower the price, get the air conditioner fixed, and load a little trailer with too much furniture and winter clothing. She'd be home in a week, in time to celebrate her birthday.

That same day, a Wednesday, Peggy Dobbyns, the secretary in the school maintenance garage in Brownwood, Texas, got a long-distance call from Jim Jenkins. He was calling from Minnesota about some pension money due him. He was almost jolly. Jim said he had found a farm. He'd bought some used machinery. Now he and Steve were waiting on credit checks from a local bank. They needed to borrow more money to buy cattle. Jim had even gotten some bad teeth pulled, so that poison was gone from his system. He sounded very happy.

Toby didn't travel that weekend. A couple of times he had driven the 225 miles to Fargo to visit Lynnette and his girls. Karen said she understood. Things hadn't gone too badly on those visits. Lynnette knew about Karen by then. She was hurt, but she said one affair didn't need to wreck a marriage and his little girls needed him and they did all have a good time together now. She made Toby a big dinner and said why didn't he postpone the final divorce proceedings a while, and Toby said he would. Lynnette made an appointment for marriage counseling.

Back in Minnesota days later, Toby showed Karen a copy of a letter to Lynnette that said he wasn't going to delay the divorce. Their marriage had been over for a long time and she shouldn't blame Karen. Toby told Karen he'd already mailed the letter to Lynnette.

That weekend late in September Toby stayed at Karen's house. She had thought she'd seen a pickup truck hanging around their place. Lynnette didn't have a white pickup truck, did she? Toby let Karen shoot the big shotgun out back a couple of times. He cleaned it; hunting season was coming. Toby started teaching Casey about the little shotgun he'd bought for him, emphasizing safety to the boy and his mom. "You're not beneficiary on my life insurance yet," he joked to Karen. Toby played with Nugget, his golden retriever, who really retrieved. Toby and that beautiful animal played practically every day. They wrestled and Toby threw sticks and Nugget fetched them. The dog was always so excited when Toby came home. Toby also put up the storm windows on Karen's house.

On Tuesday, September 27, Susan Blythe was due back in Ruthton. But she was running a day late. She'd be back late Wednesday evening instead. In Ruthton at the welding shop that morning, Swen Borresen got a call from some guy over in Long Prairie. Jim Jenkins wanted to buy some cattle. He had given Swen's name as a credit reference. Swen said, accurately, that Jim had always been right prompt in paying his bills there. But, said the welder, the fellow really ought to check with the local bank, the Buffalo Ridge State Bank.

Shortly after, the phone rang up the street at the bank. Toby Thulin took the call. Some guy in Long Prairie wanted to check the credit of a man named James L. Jenkins. Toby got the file.

JENKINS, James L. Well, said Toby, Buffalo Ridge couldn't give him a very good rating. The guy had declared bankruptcy back in 1980. Left a big debt at the bank. In fact, the bank still hadn't sold off the Jenkins house.

Thanks very much, said the caller.

A little while later the phone rang again. Rudy answered. It was a cattle dealer over in Long Prairie. A man named Jenkins, James Jenkins, wanted to get some cattle from him and when he'd called a little while ago, a Mr. Thulin had said Buffalo Ridge had had bankruptcy problems with this Jenkins guy. But now, the caller said, his would-be customer claimed never to have done any business with the Buffalo Ridge State Bank. There must be a different Jenkins, right?

Well, said Rudy, what's the social security number of your Jenkins?

The man read it off the loan application.

That's the same son of a bitch, said Rudy. And I wouldn't loan him the time of day. He is worthless. Or words to that effect.

The caller thanked him and hung up.

That night at dinner at Karen's house Toby was

shaking his head at the nerve of some people. A guy named Jenkins, he said, had stuck the bank with a bad debt three or four years before. Then he declared bankruptcy, something to do with his divorce. He still hadn't paid any of it back, but he was going all over this corner of the state looking for a loan and listing Buffalo Ridge as a credit reference. The bank had gotten a couple of calls checking on him. Some people, huh? Pass the ketchup.

Wednesday after lunch Rudy took a call in his office. Somebody had seen the For Sale sign out at the Jenkins place. He sounded like a farmer. He wanted to see it right away. Well, sure, fine, yeah, be happy to, but it'd have to be tomorrow. How about 10:00 A.M.? Rudy was in a hurry. One of his potential problem loans had been paid back, in full. Super news! Ninety thousand bucks! Wouldn't he have a lot to tell Susan that night? With the appointment made for the next morning, Rudy raced off to Pipestone to deposit the check. He picked up Rolph from school and zipped home to get the cake and things ready for Susan's surprise birthday party.

Susan phoned from dinner in Sioux Falls. Running late. But only seventy-five miles to go. Texas was a long way away. It would be good to get home.

When Lyle Landgren cruised down the main street that evening, he spotted Toby unloading the bank's lawn mower from a pickup truck. Toby was a little annoyed. He had mowed Karen's lawn and was bringing the machine back when a state trooper stopped him. It seems the truck had been repossessed, but no one except the trooper had noticed that the registration sticker on the license plate had expired. Toby got a ticket. Lyle laughed and took Toby with him on patrol for a few minutes. Toby and Karen wanted the policeman and his wife to come for dinner Saturday. They wanted company when they each quit smoking that day. Toby and Karen also had an announcement to make, and Lyle, who had heard some of Toby's horror stories about Lynnette, guessed what that meant.

He dropped Toby off back at the bank and cruised toward the Blythes'. Ah, Susan had returned. He waved. Didn't Rudy look happy? Susan wasn't such a bad egg, he thought, but she'd always be a newcomer in a place like this. Judging by the trouble she was having backing up that trailer, she wouldn't make a very good truckdriver either. Elsewhere in Ruthton, Lyle found everything quiet. Darkness was coming earlier these days.

The sun was also rising a little later. So were the Blythes on September 29, 1983. They were tired and behind schedule. Rudy weighed himself: 251 pounds, down from the 264 of just two weeks before, but up a pound from yesterday. Musta been the cake. Susan seemed in something of a rush; she hadn't even made real coffee. They were out of bottled water. But it was like Rudy had told Dick Ness; Susan was not only his wife, she was his best friend. And Rolph was himself. Gee, it was good to have everyone back together again. Things would improve at the bank. Rudy was sure of it. Meanwhile, here was Rolph, his big Tigerboy, ready for school. What a champ! Rudy and his son hugged each other hard that morning and they clung to each other a little longer than usual, as males do when they're trying to say something but can't.

Susan hadn't done the up-in-the-morning-get-Rolph-eighteen-miles-to-school-in-Pipestone-by-eight-thirty routine in a while. She was disorganized and slow, but she got to Pipestone all right and looked up as she passed by the First National Bank's digital time sign: 8:26. Uh-oh, she thought, it's going to be close. She'd promised to bring the car to Rudy by nine.

On her way back to Ruthton, Susan passed the Hartsons, Dick and Marlene, going the other way to open their business. Hartson lived across the road from the Jenkins place. Of course, like many farmers he worked off the land, too, to increase his family's income. As a neighbor, he'd kind of kept an eye on the Jenkins property for the Blythes, even alerting Susan once when he saw unfamiliar men cutting trees there. In fact, just a month before, Dick had seen a couple of men in a white pickup on the Jenkins property. He'd phoned Rudy—the Hartsons had the nearest phone to that farm—but by the time the banker got out there, the strangers had left.

When Susan arrived back in Ruthton with the bank car, she was worried that Rudy would be angry. It was a few minutes before nine. She was right on time to exchange vehicles. But Rudy's car was not in its usual spot by the bank. Nor was it by the house. He must have gone on out to the Jenkins place early. It would be just like him to go out there an hour early to make sure everything was ready for the appointment.

So Susan decided, on the spur of the moment, to drive out to the Jenkins house on County Road 7, lucky number 7. In a way it would be too bad if that farm did sell. Susan cherished the peace and calm she sometimes sought there, so far from the phones and bother and raging ignorance of life elsewhere. She had suggested the bank donate the tiny farm to the county as a refuge for abused wives. But the tax status wasn't just right or something and, as someone rightly pointed out, what frightened, abused farmwife would flee an enraged husband in an isolated farmhouse and seek to hide in another isolated farmhouse next to a concealing woods so far from help?

It was cold with a misting rain as her car crunched its way slowly up the long, curving driveway to the Jenkins house. The cable, which was strung across the drive as a gate, was down. Rudy was there. Or somebody was. You know, it was a little strange that Rudy would agree to first meet a new customer, whoever it was, out at a house for sale. Usually they met in town to talk in the office and then went out together. Might ask Rudy about that, but she guessed he was so eager to unload the place that he'd meet anyone anywhere to sell it.

With the backyard hidden from the road, Susan didn't see her husband's car for a minute. Then, oh, there it was! The big, heavy green Ambassador station wagon that had made so many trips, usually heavily laden, back and forth to Texas, and back and forth past this place too, and to all the farm auctions. But whose vehicle was that? Rudy's car was pulled up, nose to nose, to a pickup truck. A white pickup truck.

Susan pulled in behind the station wagon in the farmyard and turned off the motor. Silence. Absolute silence. Usually that's what Susan liked about the place. Its silence and its warmth when the prairie sun beamed down through the trees. But with the fog it seemed eerie somehow. People were supposed to be here today. Even the birds were quiet this morning. Must be the rain.

Susan didn't get out of the car. She surveyed the empty homestead through the drizzle. She didn't like that house; the property, yes, but not that bleak house so full of hate. The open holes in the walls and floors. The holes were still there like gaping wounds that refused to heal. Even in the summer, the house's radiators seemed unnaturally cold to the touch, like death. That forecast last night had been right about the changing weather. Susan waited, a little concerned. No one was there. Or no one was in sight. Somehow Susan didn't want to shatter the quiet with a shout.

Just then, she saw some movement in the house. Someone—no, two people—had passed by the darkened, dirty stairs window inside, descending slowly from the second floor toward the front door.

Susan didn't approach the house. She waited in the car. When the door opened, it was Rudy; even in the mist and rain she could spot him in that bright yellow rain slicker. She'd wanted to get him one for his birthday, but the catalogue store had been out, so Rudy bought his own birthday present from J.C. Penney. Oh, and there was Toby. What was Toby doing here? She hoped they wouldn't talk about having dinner together with Karen. Rudy had suggested the four of them dine out. Privately, Susan said it didn't look right, seeming to condone Toby's living with someone before his divorce.

There wasn't much time to chat. Big Rudy was visibly annoyed. Actually, he was past annoyance and moving well into anger. Someone was on his property when they shouldn't be. Rudy thought that it might be that son of a bitch Jenkins. Rudy had referred to him so often like that around the bank that son of a bitch seemed like Jenkins's first name. It looked as if Jenkins was there. He liked white pickups—he had the money for them, thought Rudy, but not for his old debts. The truck had Texas license plates, or one of them anyway; the front plate was missing for some reason. And there was a Jenkins checkbook on the dashboard.

Damn, this pissed Rudy off. Can you imagine what the potential buyer would think when he got there in just, what, fifty-three minutes? A fruitcake former owner hanging around the property nearly four years after he walked away from it? Hard enough to sell this lot when no one around had any money. Rudy wandered north alongside the garage, calling out into the woods.

"Hello. Anyone there?" There was no answer.

Susan was a little concerned as she stood with Toby by the car, so she kept an eye on her husband. She was also a little cold, so she pulled some sweat pants on over her shorts. Rudy had walked past the garage several feet and was nearing the edge of the woods. There is a slight incline there. So big Rudy looked shorter, standing in the tall, soggy, dying weeds and looking around, calling out. There was no answer.

Rudy, his jaw set, came striding back toward the vehicles. He was opening his mouth to speak when Susan heard a funny metal noise back in the woods. "What was that?" she said. "Did you hear that, Rudy?"

Rudy didn't answer Susan's question. And that sent a piercing chill through her body despite the warm clothes.

Maybe someone feeling less financial pressures, someone smaller who hadn't wrestled and played football and been big all his life, someone with more awareness of sudden midwestern storms, would have been frightened, would have sought shelter sooner. But Rudy was from the East, a take-charge guy who could be very stubborn, like an old farmer. Especially when it came to dreams.

"Go get the sheriff," Rudy told Susan. Then he turned his head toward the woods and in a loud voice said, "GO GET THE SHERIFF AND TELL HIM WE'VE GOT TRESPASSERS ON THE OLD JENKINS PROPERTY."

"The Pipestone sheriff?" asked Susan.

"No," said Rudy, "we're in Lincoln County here. Go to the Tyler Town Hall and phone Abe Thompson from there."

Susan quickly jumped back into the bank's car. This was official bank business now. She hurriedly made a U-turn toward the barn and headed down the drive. By then, Rudy was walking back toward the garage, and an unconcerned Toby was headed off down the path toward other farm buildings like an inexperienced point man on patrol in Vietnam.

In her mounting haste, Susan spun the car tires on the drive. Rudy's not answering her question had scared Susan. She turned to the north on County Road 7 and accelerated quickly, only to slow slightly at the Hartsons' drive. Try their phone? No, wait, not home. She'd seen them in Pipestone. Susan pressed the accelerator to the floor and pulled out to pass a slow-moving pickup. Damn farmer. Soon she would be doing seventy miles an hour and worried about making the corners. Susan looked back at the Jenkins house out the side window, trying to spot her husband. But the protective trees had already enveloped the scene.

The first shot blew melon-sized shatter marks in Rudy's windshield. Glass went flying inside the car and into Toby's hair. What the—? He ducked down in the passenger seat, terrified. The next missile from the woods came through the little triangular vent window, not the glass but the rubber stripping around it, creasing the chrome but staying whole.

The bullet went straight to Toby's throat, right through the middle beneath his chin. It was a very neat, small red hole about the size of a pencil eraser. In the back it was much larger. The .30-caliber slug shattered inside the veteran's neck and exploded out the back, tearing away the carotid artery and the entire spinal column. Toby's head jerked. His eyes flew wide open. The body convulsed violently once. Then, it slowly began tilting to the right against the open door. Blood spurted on the handle and ran down the door to drip in the dirt. *Bam!* A third bullet slammed into the door right where the head was. The metal stopped its flight. Toby fell backward until he hit the ground with a thud. His feet, clad in his new brown loafers, not his combat boots, were in the car. His face was looking up at the clouds, the eyes now half closed. Slowly, a thin, steady stream of blood oozed out the front of his throat and trickled down into his right ear.

Rudy had been hit by that first shot. The windshield sheared off the bullet's copper cover and slowed its velocity. It was the least lethal shot of the day. But for investigators it would be most helpful later. The nearly spent piece of lead flew into Rudy's lower back above the buttocks. He was crouching as he left the driver's seat. The bullet might have been enough to knock him down, but Rudy never noticed. He was up and running, past the side door of the Jenkins house, along the little cement walk, across the yard toward the road. He knew! The Hartsons! Had a phone. Long way.

God, what was happening? Who was shooting? Rudy's cordovan wing-tip shoes were slipping badly on the wet grass and bumpy, unfamiliar terrain. Rudy had a lot of weight to move. His mind wanted to move the body faster, much faster, than its condition would allow. He was puffing already, cutting across the overgrown lawn toward that other house so very far away just across the road.

A pair of booted feet moved swiftly away from the woods through the tall weeds of the empty farmyard. They went to the car, the passenger's side, and stood there for a moment. So that's what death looks like. Welcome to the farm, fucker. And who was that anyway?

Then the boots moved quickly around the back of the green station wagon, then faster past the side door of the house, and faster still along the little cement walk, clomp, clomp, clomp, and into the middle of the big front yard. Now, where'd he try to go. Oh, there he was, the sucker, lumbering through the ditch, heading for its broad, steep banks, probably trying for the Hartsons' place. How does it feel now, Mr. Banker, to be outside your own office with the fancy furniture and all them files and your money tricks? You're not in charge out here.

Rudy had been an army cashier, not a commando. There were trees on three sides of the yard, but Rudy was running out in the wide open spaces with his big body wrapped in his bright yellow rain slicker.

The boots stood carefully spread apart. Rudy was bent over, still running but now desperately trying to hide amid the flimsy hip-high weeds. The man in the boots raised his arms and tilted his head ever so slightly. No rush. The target's broad waistline sat right atop the sight. Rudy Blythe, the banker, was about to be foreclosed.

Take a breath. Let some out. Hold it. Squeeze. Squeeze. Perfect. The explosions surprised even the man in the boots.

Propelled by the hot gases, the two bullets tore across the yard almost simultaneously. Eighty-four feet later they struck just above the beltline on Rudy's right side barely a half inch apart. But because the forty-two-year-old father was running bent over, they tore up through the body, shattering and sending lead chunks slicing through arteries and vital organs.

Whoof! The air rushed from Rudy's lungs. The impact straightened him up. He whirled around. A third bullet ripped through his upper arm. A fourth zipped past his ear to harmlessly pierce the jacket's hood before sailing off over the empty Hartson house. Suddenly, darkness was falling. C'mon, let's hear it for the team. By the time the 251-pound body slammed down on the wet, cold ground, his face looking up, his left arm pointing toward the Jenkins house, Rudy Blythe was gone.

It was a pretty normal day. Abe Thompson was his usual self, serious, quietly friendly, saying much less than he thought. Years ago, he'd served his time in the military, back before young men figured they could refuse. Abe didn't much like army life, so the day his time was up was the day Abe got out. Came home to Minnesota where his brother would run a little newspaper and he would be a welder, a mechanic, deputy sheriff, and eventually sheriff of Lincoln County. He was reelected and reelected, and now there might be one more quiet term for the fifty-one-year-old. Then the easy life. No more badge. No more paperwork. No more farm sales; he hated them. No more teenage highway deaths; he hated them, too. They were hard to take, and harder still to figure out. As Abe walked back to his desk with a brimming cup of fresh coffee, he instinctively readjusted the pistol on his hip. It always caught on the arm of his chair when he sat down.

The moment he set the hot cup down, the phone rang. Both phones rang. The dispatcher got one. Abe took the other.

"This is Susan Blythe," said the caller quickly. She sounded scared. "We own the old Jenkins property near Ruthton. There's some trespassers there and—"

"What? Who is this?" asked the sheriff, turning in his squeaking chair. He hadn't caught the name. The woman was talking too fast. And now, what the hell, the dispatcher was talking at him urgently from the door.

Standing in the Tyler Town Hall with a borrowed phone stretched over the counter and several local residents listening, Susan knew what the sheriff wanted to hear. If she had said, "Susan Blythe, the banker's wife," he'd have known right away. Because of her fear, now her anger was rising, too. These country people—a woman's always got to belong to somebody. . . . Then Susan heard the sheriff talking to someone else there.

"What? The Jenkins place? In the ditch? Yes. Right. I've got her on the other line." Then, coating his voice with caution, back into the phone, "Mrs. Blythe, is that you? Where are you?"

"There are trespassers on the property and my husband said to call you from Tyler."

"Yes, I know," said the sheriff, reaching around to check that his holster was full. "I'm on my way."

"Should I wait for you here," asked Susan, "or at the property?"

"At the property," said the sheriff, who was already

moving toward his car. This time he put his seat belt on. Within seconds the cruiser was hissing down a quiet, wet Highway 75 at eighty miles an hour. It was the kind of unusual, urgent motion in the countryside that told onlookers by their mailboxes that something was very wrong somewhere.

Susan was puzzled. And she wanted to stay that way. If she let her mind think too much about why someone else would be calling the sheriff's office about the Jenkins place, where her husband was, she would panic. It was she and Rudy and little Rolph out here, alone, surrounded by violent strangers. If anything ever happened to Rudy . . . She methodically put her mind to other things, driving slowly back down the county road. She was in no hurry this time. There might be only a few minutes left to savor. She even noticed with a smile an old house that she and her husband had admired. Rudy had a way with houses, all right. He could see the outside and describe the interior layout. For some reason, Susan thought about that first date in Philadelphia.

Then she saw the truck, and began, unwillingly, to understand. It was parked by the Jenkins driveway, her driveway, in the middle of the road with its tail-lights flashing. A man was standing in the road, waving her over.

"Ma'am," he said through the open driver's window. "You can't go in there."

"Why not?" replied Susan, who kept inching the car along. "Who says?"

"You just can't, ma'am," he said, walking along, fumbling for his wallet. Then he flashed a badge. It was Paul Bartz. He said he was a part-time policeman. He reached in and turned Susan's steering wheel to the roadside. Susan got out to talk.

"You can't go in there, ma'am," he said, physically blocking the way. Oh, my God, thought Susan, it's a trick. These farmers have got Rudy, and I walked right into the trap too. She hadn't looked closely at the badge. "Okay," she said, stopping her resistance and turning around. "I'm okay." She started to walk back to the car. The man let go.

But Susan wouldn't face their rural reality without a trick of her own. She whirled around and flung her fist at the man's face. It bounced off his chest and she began flailing.

"Stop it!" the man yelled. "Your husband is dead in the ditch!"

Susan froze.

There, about twenty steps away, partially hidden in the matted grass, was the yellow jacket, still, very still. If you confront things that are wrong, Susan had always believed, anything can be fixed. But this, she knew instantly in her heart, this could not be fixed. As Susan walked toward the yellow jacket, she saw the khaki pants, too. They were wet now from the rain. He must be cold. And his good wing-tip shoes were muddy. Now he'd really have to polish them. And his face, where was Rudy's face? Someone had put a gray sweat shirt over it. The fabric was dark now with all the rain. If only she'd gone to church more, Susan thought, maybe this wouldn't have happened.

Susan knelt down beside her husband and lifted the covering. His mouth was open. He was a strange color, a bad color. And his jaw was pulled back funny. When she'd left just a little while ago, Rudy had been full of life. Now he was just an empty shell. Where did he go? What would she do? What could she do? Susan began to focus on the little things, the big ones being unthinkable. She began arranging Rudy's clothes, pulling his jacket down and tucking his shirt back in. He was warm. That was a good sign, maybe. And the arm wound didn't look all that bad.

When the volunteer medical crews arrived, a sobbing Susan was holding Rudy's hand, the one with the big Mason ring. "He was a very good man, you know?" she said.

"Yes, ma'am," they said, standing around the Blythe victims in a semicircle, their equipment bags ready.

"Did you know him?"

"No, ma'am. We didn't. Can we have a look, please?"

"No!" shouted Susan. "Get away! Haven't you all done enough? Leave him alone. Leave us alone. He's dead, can't you see that? They've killed him. That goddamned Jenkins." And she rubbed her husband's face.

Sheriff Thompson arrived then, siren wailing and lights flashing. "Why did they hate us so?" Susan asked him. Pipestone County's Deputy Sheriff Lyle Landgren came. And LeRoy Burch, Ruthton's mayor, who was a paramedic and Rudy's friend. But a sobbing, swearing Susan would let none of them near her husband. The big guy seemed so vulnerable now. He had protected her in that damned farmyard, probably saved Susan's life by sending her away. Now it was just Susan here in enemy territory. Just her and, oh, my God, Rolph! He's in school in Pipestone. He mustn't hear it from anyone but Susan.

Her fast, frightened thoughts were interrupted by the police and medics. "What happened?" they asked Susan.

"I don't know," she said through tears. "That son of a bitch Jenkins. Ask Toby. He was up there too." She motioned toward the house.

Susan didn't notice some officers break away from the gathering roadside crowd and creep up the driveway through the trees, their guns drawn. Nor did she see them walk down the driveway a few minutes later, their guns away, their heads down. Lyle broke off then and headed into town, to the bank. Paul Bartz stayed at the scene. His firsthand details would prove crucial. He had been driving south on the county road, had seen a small car flash by going north, and then, as he passed the old Jenkins place, thought he saw a bright color in the ditch, where bright colors do not belong in late September. As he turned his truck around on the narrow road, a white pickup roared down the Jenkins drive and raced by him to the south. An older man was driving, wearing a stocking cap.

"We just want to do our job," the paramedic was pleading with Susan.

"I know," said Susan, who was beginning to think it didn't matter anymore. Her world was ending in this godforsaken place. One more foul indignity wouldn't matter to the dead. They began examining the body, squeezing the limp arm for a nonexistent blood pressure, prying an eye open to check the pupils, even trying chest massage. But with each push on Rudy's big chest, blood spurted up into the oxygen mask. The medics pretended not to notice. Susan, wincing, moved her knee against Rudy's hip to quiet a gurgling sound there. She was going numb.

"He's not usually this overweight," she said to the bustling medics, who were not listening. "He's gone on a diet, you know."

When Lee Bush drove up to the developing traffic jam by the ditch, the deputy was disturbed. "Lee," he said, "we've got one helluva mess here. We've got two dead bankers and a hysterical wife."

Lee knelt with his friend's wife in the ditch. "Susan," he said, putting his arms around her for several minutes. He was the only person there she would call a friend. "Susan," said Lee, "Susan, we should go." He told her that the bodies—yes, Toby's dead too—would be there for hours while the police took pictures and sought clues to what happened. By then Lee, who had talked with Rudy by phone less than an hour before, wondered how close he had come to being invited to the farmhouse rendezvous to help close a deal.

"Well," said the new widow, standing up in the wet ditch, "I've got things I must do."

The news of the two deaths spread quickly through the midwestern countryside and then, somewhat more slowly, seeped into the general news stream of a preoccupied nation: PILOTS SET STRIKE AT CONTINENTAL; BRAZIL'S DEBTS CAST A SHADOW; CONGRESS ALLOWS MARINES TO STAY IN BEIRUT; BANKS FREE TO OFFER HIGHER INTEREST ON SAVINGS—DAYS OF EASY BANKING OVER. *In Minnesota, police said the president of a rural bank and his chief loan officer were shot to death today after they were lured to a farm the bank claimed in a foreclosure proceeding four years ago. And in sports, Mike Warren, a rookie for the Oakland A's, pitched a no-hitter today, defeating the Chicago White Sox three to nothing. The weather after these messages . . .*

Lynn Carpenter had always liked Rudy as a boss.

She always worked hard, so he trusted her. She and three other women in the Buffalo Ridge State Bank were at the new computer, which was fast becoming obsolete, posting quarterly interest payments to savings accounts, when Lyle Landgren walked in with a shotgun and said the bank was closing for the day. He took the four women into a back room and broke the news. There were gasps and sobs and stares. An automobile accident? No, murder. Murder? Lyle said no one could touch Rudy's or Toby's office until the boys from the state Bureau of Criminal Apprehension got there later. He also wanted everyone to stay away from the windows. There were two men in a white pickup truck—they think it's Jim Jenkins and that boy of his, the one that always dresses like a soldier—seen leaving the farm after the shooting. They're still out there somewhere. There were reports they had grenades and maybe a grudge hit list.

By 10:05 that morning Special Agent Robert A. Berg of the state BCA was speeding toward the scene of a reported double homicide near Ruthton, an investigation that would take over his personal life, straining his marriage a little. Mike Cable was in court up in Ivanhoe, handling the last routine property theft prosecution before the biggest case of his career blew Lincoln County's yearly law enforcement budget. DuWayne Schroeder and Van Herringer and Toby's other buddies from the drill team up in Ada, Minnesota, or down in Sibley, Iowa, heard the news on the radio or from a storekeeper who heard it on the radio. They couldn't understand how a seasoned war veteran like Toby could walk into an obvious ambush.

Toby's wife, 225 miles away in Fargo, North Dakota, wouldn't get the news for several hours. Lynnette was in school, and Lyle and others had difficulty tracking down her new minister to break the news.

Karen Rider had been taking a shower at 9:10, mentally preparing for her afternoon class of local retarded students. If Toby hadn't borrowed the bank's repossessed pickup the previous night, Karen would have driven him to work in Ruthton and they would have lingered over a casual breakfast in the cafe there.

But Toby wanted to tell Rudy about the truck's registration problem, maybe get that settled first thing. So he'd driven to the bank alone, and earlier than usual. Now Karen would take her daughter's forgotten gym shorts over to school and then zip down to Ruthton to surprise Toby for lunch.

When Karen walked up to the bank door that morning, she thought it was a little funny. The door was locked and the sign said CLOSED TODAY. Toby hadn't said anything about that. Was this a Minnesota state holiday or something? As she walked away, Karen heard a furious rapping on the glass. It was Lynn Carpenter motioning her to come in.

"I'll wait for Toby over there," Karen said, pointing to the cafe.

"No," said Lynn, "please come in here."

The lights were low in the bank. They were out altogether in Toby's and Rudy's offices. What was the deputy sheriff doing here? "Where's Toby?" Karen asked.

"Please come back here," Lyle said.

"What is it?" Karen demanded, louder, her hands wringing the bottom of the sweat shirt, one of Toby's, she was wearing. "Where's Toby? Where's Rudy?"

"I'm sorry, Karen," said Lyle. "There's been an accident. They're not here."

"Where are they?" asked Karen, hoping against hope and knowing full well there was but one hospital nearby. "What hospital is it?"

"They're dead, Karen."

Richard and Beverly Grabenkort were in their Dallas living room all packed and ready to go. Two weeks alone together in Europe. The children would be cared for. Everything was arranged, even someone to check now and then on the Blythes' empty house next door, and water the plants. The pool was turned off.

Three days before, Richard had helped load the trailer for Susan to haul back to Ruthton. Now, just before they left for the airport a half-hour's drive away, Richard called Susan at the Minnesota bank to make sure she had arrived home safely. When he had asked for Rudy or Susan, the bank secretary had said nothing.

"Hello? Hello?" said Richard, but the secretary was handing the phone to a man standing nearby.

"This is a deputy sheriff," said the voice from far away. "Mr. Blythe has been murdered. He's been shot."

Richard's face turned white. He hung up, walked into the living room, and as his wife looked up, smiling, expecting the good news, he repeated the deputy's announcement like the blast of a shotgun. Then he walked absently out the door toward the Blythe house.

Beverly sat stunned for a moment, the obsolete smile frozen on her face. Because she couldn't, wouldn't think of jolly Rudy, by his swimming pool in his plaid shorts with his camera, now lying somewhere very still with neat red holes in his body, she focused instead on her husband's announcement. What is it about men, she wondered, that they have to get these emotional things out so fast? Is it because they're too hot to handle? But it wasn't a question.

Dick Ness was out in the shed working with a couple of friends. Ever since he'd come back from that fishing trip up north with Rudy, his pickup truck had been misbehaving. Catherine Ness was in the kitchen when the phone rang. It was a friend over in Worthington calling to see if they were all right. He'd heard something about a double homicide in Ruthton, and somebody from Texas in a white pickup truck was wanted.

When Catherine took the news to the shed, all the men agreed that it was most likely old Jimmy Lee Jenkins. They figured the victims were his ex-wife Darlene and her boyfriend, Louis Taveirne, which was too bad but they didn't know them personally. However, when Catherine returned to the kitchen, the phone was ringing again. It was her thirteen-year-old daughter, Lyrae, in tears at school, where there was chaos at the news of the murders. Before the mother could calm the distraught girl, the youngster announced who had really died.

Rolph knew something was going on, but he thought

he'd misbehaved. The principal, Mr. Zorich, had taken him from class in the middle of the morning and kept him by himself in the school auditorium. Now he sent the eleven-year-old upstairs to get his coat and books. And there was his mother waiting downstairs by the front door. She looked as if she'd been crying. Maybe Rolph wasn't in some kind of trouble after all.

"There's been a terrible accident," Susan said, kneeling down on the wet cement by her son. She had changed out of her bloodstained sweat suit. Her eyes were swollen. She felt herself swimming in a vast ocean with no land in sight. She was squeezing Rolph's hand very tightly, as much to get strength as to give it. She wanted to make the awful news easier somehow, but she couldn't think what to say. It didn't come out right—and Rolph remembered it differently later anyway.

"Daddy's dead," she said.

There was a long pause, perhaps two seconds. Then Rolph wrenched away violently from the messenger. "No, no, no!" he screamed. "Not my dad! Not my dad! You're lying! Not my dad!" Rolph ran away from the news across the lawn in confused circles and ended where he wanted to be, in Susan's arms.

Susan could have used some hugs herself. Now she was alone, in charge of everything. She had come back to this hated place because she loved her husband and he had asked her to and twelve hours later he was gone, forever, and she had to do everything herself, and what would she do for money and for Rolph and about running a bank?

From the ditch Susan had gone to the bank with Lee. She called Rudy's parents in Florida, but his father was at an investment club meeting. Rudy's mother was there, but she was alone except for the cleaning lady. Susan didn't want to tell her, but Dot was already suspicious. She hadn't recognized Susan's voice at first; it wasn't like her daughter-in-law to phone in the morning. There was no choice for Susan now. "Rudy's been killed," she said, "and—" The line went dead.

By the time Susan got through to Rudy's brother,

his mother had already called him. "I know," George said. Susan was feeling increasingly cut off from everybody. Sharon Fadness had arrived at the bank and that was a help. They had become much closer during the drives back and forth to Texas. Susan's anger would fade to a stolid determination in the months ahead. But for now she was still very angry and lost. And when a minister arrived and suggested perhaps that the destruction of Susan's life, her abandonment in enemy territory, and the snatching away of her husband forever was God's will, Susan told him, quite bluntly, to leave.

Susan, Sharon, and Lee had then stopped by the Blythes' house. It seemed very empty. The police did not want Susan there, not while the killers were loose. They had to go for Rolph right away, but Susan needed fresh clothes. On the way out, she grabbed her birthday card and her husband's pajamas, physical links to someone, somewhere.

Susan, Sharon, Rolph, and Lee went in Lee's car from the Pipestone school to Sharon's house in Ruthton. Rolph, in the back seat, was distraught. He kept yelling, "No, not my daddy!" and kicking the seat in front. At times, Lee thought Rolph might tip the little Ford Escort. Susan was worried too. Nothing would calm the boy, so a mile out of Pipestone they turned around to go for a sedative.

Then, suddenly, Rolph calmed down. "I'm all right," they heard him say, although he wasn't speaking to anyone in the car. He was talking to his father. "I said I wasn't going to cry anymore," Rolph recalled later. "He wouldn't like it."

For the rest of the day until they moved into a Pipestone motel, Rolph and Susan stayed, or rather hid, at Sharon's, first from the killers and then from the press, who, to the locals' surprise, flocked into town to interview anyone on the streets or in the cafe. Everybody said they were shocked about the shootings, couldn't understand how they could happen. In the city maybe, but not out here. "Who can know what these hard times do to some people?" said Duane DeBettignies, the editor. "Every man has his breaking point. But how are you going to blame a banker for bad times? He's got to run a business too." Bernard Knuth noted that Jim Jenkins was an only child. "His parents thought the sun rose and set on him," he said. Tammy Burch, the mayor's little daughter, remembered how close Rolph and Rudy Blythe were. Little Greg Martin admitted that Rolph did get a lot of teasing because he was so rich. "Rudy had tried to be helpful to some of these farmers," Gary Lindahl said from behind his cafe's busy cash register, "but maybe he let people borrow when they shouldn't have, you know? And lately he was getting a little worried."

Many of those interviewed mentioned the good things Rudy had done for the town. A few, in a lower voice, allowed as how the bankers had been leaning on some folks pretty hard recently. But, you know, killing somebody is a mighty extreme way of paying off a debt. Yeah, well, banking's a business like any other. That's right. Those borrowers did have an obligation to repay, but bankers, well, they're a different kind of person, you know what I mean? Oh, and you won't use my name, will ya? On that last part anyway. Bankers read newspapers.

A few people gathered at Sharon's house throughout the day, bringing food and solace and wondering about Susan's behavior. She wanted no pills—must stay alert in case the police needed help. She walked about the house, talking in disjointed half sentences, believing herself remarkably coherent, and wondering when her parents' plane would ever arrive. But no one questioned Susan and no one interrupted her.

Rolph, clutching Rufus, struck everyone as quietly mature on That Day, as September 29 came to be known among family and friends. From a corner, the boy watched the morning's events change the adults, especially his mother. She was acting very strangely. Rolph was certain he was losing her, too. Sometimes, with his arm around Rufus, who sat loyally by his side, Rolph gazed at the TV set, at one point spotting minicam coverage of an ambulance crew wheeling his father's draped body from the ditch. That was before adults in the next room leaped to change the channel.

Rolph did a lot of thinking that day. He wanted to know what happened out there at the place he had always associated with quiet, happy times. He remembered how his father always seemed to be thinking. The boy had long wondered what wonderful, big thoughts were flowing through the big man's mind when the son watched the father sitting in the living room alone or pausing to look far away from his bank desk. Rolph had never really asked his father what he was thinking, but the boy would always wonder. He had thought there was plenty of time for asking.

Karen's two children took Toby's death quietly and hard. Counting the painful divorce, he was the second man they had lost in their young lives. There were hot tears and then blank looks, until, sitting at a table in Lyle's house, they heard the deputy sheriff describe the Jenkins men to Karen. At the description of Steve— shaved head, tattoos, camouflage clothes, machete on his leg—little Casey blanched. He had seen that man walking by their house several times in the last week. Lyle looked quickly at Karen, who soon arranged for the children to visit Iowa for a few days.

Outside the victims' small circle of friends and families, and the investigators, life would go on pretty much as usual in the coming days. The leaves needed raking. The familiar big red farm trucks, their high wooden sides reinstalled for the bulging burdens of autumn, lumbered up and down the dirt roads, hauling their golden cargoes of grain to rippled steel storage bins or to the tall grain elevators.

The Ruthton bank reopened on Friday. The Ruthton Vikings football game that night went on as scheduled against Lake Benton's Bobcats. Ruthton lost, 22–14, on an intercepted pass and a touchdown in the final seven seconds, a major lapse by the defense that the coach wouldn't criticize until Monday practice after the loss had healed some. For the outsiders wandering Ruthton's streets those first couple of days, the old yellow station wagon parked outside the Standard Oil station meant nothing. Its repairs were complete, but Toby Thulin had never returned to pick it up.

The manhunt began immediately, though it was severely hampered at first by the fog and rain and low-hanging clouds. From the Jenkins farmhouse, Lyle Landgren had radioed to surrounding counties a description of the men, the truck, and the rear license plate, which trusty Paul Bartz had noted as the white truck sped away—black letters and numbers on a white plate, it looked like. Sounded like Texas. But the main radio tower wasn't working right that morning, so the garbled message didn't get very far at first. Only a few deputies were looking for a Texas pickup, something about speeding in Lincoln County.

Soon after the shooting a middle-age man and a teen-age boy dressed like a commando bought a hundred rounds of hollow-point, .30-caliber carbine ammunition from Harvey's Trading Post in Luverne, Minnesota, forty-three miles south of the ambush site. The men said they were going hunting. The same pair bought some .410 shotgun ammunition and a flashlight at a local hardware store. Both purchases were paid for with checks drawn on the account of a James L. Jenkins.

Soon after, Pipestone County Deputy Sheriff Ron McClure saw a 1978 white Chevrolet pickup coming at him on Highway 75. It probably wasn't the wanted one; this one was headed north, back toward Lincoln County. But as they approached, McClure noticed that the front license plate was missing. He stopped, spun around, and caught up with the truck. The rear plate was from Texas, KW3618. Officers would soon learn that plate number was registered to James L. Jenkins in Brownwood, Texas.

Deputy McClure followed the truck quietly without a siren or flashing lights. The truck wasn't speeding. The deputy saw two men—one of them, the driver, wore a tightly-rolled stocking cap—look back at him occasionally. The truck seemed to be heading for Hardwick. McClure dropped back a little. Then the truck turned off on a dirt road and stopped. The deputy saw the passenger get out with a stick and look back at the road and then—Jesus, it was a rifle! Aimed at his car! McClure sped past the dirt road. He thought he heard shots.

He stopped. The passenger got back in the truck. It drove away. Just as the truck went over a little hill, McClure saw the brake lights go on. Ah-hah, he thought, they're going to wait there to ambush me. He radioed for help and waited, but by the time the cars and helicopter arrived, the truck had disappeared. McClure found a couple of .30-caliber carbine shells in the dirt.

For the state detectives who thronged Ruthton, unraveling murder and mayhem is old stuff, Monday through Saturday, and, in the high-profile Blythe case, Monday through Monday. For Lincoln County's Sheriff Thompson, however, much of this was brand new. "I've handled assaults and rapes and suicides and just about everything," he said in his Ivanhoe office right next to the courthouse, "but I tell you I'm learning a whole lot about murder real quick."

The sheriff knew Jenkins, of course. It's a country sheriff's job to know everybody thereabouts, and not just for the votes. The folks he doesn't know out there, he should suspect. Outsiders—unfamiliar faces or trucks or out-of-county license plates—think different, break the rhythm, cause trouble. Abe knew Jenkins pretty well. He'd served him a couple of times for bouncing checks or not making a proper payment. Jenkins was a loner, all right, which is not unusual in the countryside. Not unusual in the city either, except there's too many other people to notice. "A likeable sort," Thompson said of Jenkins. "Nothing unusual. No trouble."

Abe knew he was up against a formidable fugitive. "He knows this country real well," said the sheriff, "all the little back roads and groves." What the sheriff didn't say in his meetings with the press was that he knew the countryside pretty damned well himself. Abe said he believed the killer or killers were still in the area. The FBI kept in touch in case it became an interstate crime.

The sheriff knew something of the psychology of the hunted. While the BCA boys quietly and methodically went about the business of vacuuming cars and clothes for fragments, sifting weeds for spent shells, prying into lives for personal details, and mining minds for motives, the sheriff would put a little public heat on the fugitives, make them feel alone, isolated, surrounded, and, like all prey, frightened. He wanted the Jenkins pair to feel they had nowhere to go. That could help run them down.

Abe was working on getting Jim Jenkins's full-face photo from the Texas Department of Motor Vehicles. He'd already gotten one of Steve, standing in his grandparents' kitchen looking goofy in army clothes, with his eyes closed and his tattooed arms holding up a couple of fresh-caught fish. Later, those pictures would go out to a photo-hungry group of journalists for transmission all over the region, and maybe even the country. Abe also got the grandparents, Clayton and Nina Jenkins, to film a short emotional TV appeal to their son and grandson to turn themselves in. "We love you," the grandma said.

The sheriff repeated for countless tape recorders a description of the hunted pickup truck, carefully leaving out one or two minor details to help him winnow the hundreds of eager responses he was to receive in the coming days. "At first," he said, "a lot of these citizen calls sound very promising. They are certain they have seen this pair. Then you start to check them out and they crumble in your hands. A woman up near the Twin Cities saw the white truck, the Texas plates and all, but she was sure it was a small truck. We're looking for a full-size one."

Carrying tape recorders, and with pistols under their coats on their belts, the plainclothesmen, with backup from some sheriff's deputies in surrounding counties, made their thorough rounds of the fugitives' friends, the places the father and son frequented, and the places they might have frequented. Police visits to Luverne stores came up with two checks signed by James L. Jenkins. The clerks on duty remembered what they looked like. They were sullen but, no, not nervous. They bought ammunition. The police seized the checks; it didn't matter—there wasn't enough money in the account to cover them anyway. Now authorities knew how the men were armed. But a big flashlight.

That was intriguing. What would Jenkins want that for?

In the newly leased Jenkins farmhouse near Hardwick, searchers found food, another gun, some instructions for making homemade bombs, and a telephone number with a 214 area code. When they called that number, no one answered. Officials at the Texas telephone company said the phone was on the north side of Dallas, Texas, on Meadow Haven Drive, under the name of Blythe, Rudolph H.

BCA Agent Dennis Sigafoos taped his interview with Darlene Jenkins. She said her ex-husband hardly ever did any shooting. He'd had an old shotgun years before, but he'd gotten rid of that. Steve had a rifle, she thought, an awful-looking army thing with a military sling. But that was nothing. Steve had liked military things since, oh, golly, since he was little.

BCA Agent Michael Cummings poked around Rudy's bank office. He'd found the Jenkins file easily enough. The farmer had sold mortgaged property, then declared bankruptcy. But what was this back here by the phone on the message spindle? A message about a phone call from a Ron Anderson and something about a 10:00 A.M. meeting.

Does anybody in the bank know a Ron Anderson? Pause. No, there's no Ron Anderson around here.

At Clayton Jenkins's place, the couple cooperated fully. They perhaps thought or hoped there was some awful misunderstanding that would clear everything up. Did Jim or Steve do much shooting? they were asked. No. Well, Jim didn't, but Steve did. He used to go out back for hours at a time. Just by himself, you know. That was when the agents found the thick, Y-shaped tree branch. Someone had put a pair of pants on it and a blue shirt. The wooden limb was pocked from hundreds of rounds of rifle fire.

The judge signed the arrest warrants that night.

The agents also began the first in a series of interviews with Susan, who remained a taut bundle of nerves—constantly working, talking, thinking lest she have a moment to confront what had really happened. Even much later, it was hard for her to speak of

Rudy's murder—she came to call it The Ditch—without imposing on herself an icy impersonal tone of control that would deeply worry prosecutors, who knew more than Susan about the emotions of juries. Susan was so eager to help the detectives—some of them saw it more as an obsession—that she would call them often with a freshly remembered detail or a suggestion for an investigative lead and then want to hear progress reports. They understood, bereaved or enraged spouses being a common encounter in their line of work, but it was distracting. Susan was an English teacher and a city woman, not a trained investigator. They quietly arranged for Mrs. Blythe to get some professional counseling.

Rolph's counseling came from his mother. Everyone agreed he should try to resume as many normal activities as soon as possible. Two days after That Day, even with the Jenkins pair still on the loose, the youngster went off to his beloved football practice as usual; Rolph didn't see the unmarked car parked by the field with the large man watching his every move even more closely than his coach.

That first Sunday night without a father was an important one. Susan had wanted to have a few private minutes with the body and Rolph before the public memorial service. But the BCA questioning had taken longer than planned, and she and Rolph rushed into the funeral home as others arrived. Rolph thought his father looked a little strange in the coffin. He touched his dad's hair. Susan showed Rolph how, looking down from the head of the coffin, the body looked very much like his father stretched out in his recliner chair in the living room; she didn't want Rolph building any illusions that the man about to be cremated was not really Rudy Blythe.

Inside, Susan wasn't all that calm. Even with her parents there, she felt hollow and lost. The memorial service would be the first time since Thursday that she had had to confront the physical evidence, final and unequivocal, of Rudy's death. She couldn't fully handle that yet, so she focused on details—how lovely the flowers were, there should be more, had everyone been invited, were all the arrangements in order, are there enough chairs? Try as she might, one haunting thought kept creeping into Susan's mind about her abbreviated life with big, awkward Rudy: despite all her pleas, subtle and bold, they never danced together.

That weekend, in bed, when slumber freed her mind for some moments, Susan was reliving, detail by detail, the events of The Ditch. And then, as if looking down on herself, she said, "Oh, this is really just an awful, awful dream and I'm going to wake up now to reality." She woke up and looked hopefully around the room. She was in a motel room in Pipestone. Her mother was sound asleep in the semidarkness. Rolph and her father were next door. The police were just down the hall. Rudy was not in the bed. Just his pajamas. She began to scream.

When Dick and Catherine Ness returned home to Ruthton from the memorial service, Catherine absently flicked on the telephone answering machine. They'd been so busy in recent days they had forgotten to check for calls. Catherine was on the couch listening to the tones and the brief recorded messages when she heard Rudy talking to her. He sounded so happy, so near, so alive. *Beep.* All the color slides are back from the fishing trip, he said. Why didn't they all get together for dinner and look them over? *Beep.*

Dick Ness found his wife sobbing.

Nobody told Nugget what happened to the man who fed him and brushed him and threw the sticks to be fetched, but Toby's golden retriever knew anyway. The day of the shootings Nugget began scratching at the kitchen door. When it was opened, he didn't want to go out. The dog paced and whined for days until Karen put one of Toby's coats on the floor. The dog curled up on the garment and calmed down. But he wouldn't retrieve anymore. He wouldn't eat much either. Clumps of his hair began falling out. The veterinarian said maybe it was diabetes. But no amount of medicine or attention from a string of families could fill that void. A few months later a skinny, listless Nugget went unsuspecting onto that familiar shiny table in the doctor's office. It was the last thing he saw.

6

APPOINTMENT IN PADUCAH

Sundays are quiet in rural Texas, Dr Pepper country, where cowboy hats are not worn for fashion. Even the broad, endless fields of the High Plains seem to take the day off to lie around in the wind and fading sun listening to sad Waylon Jennings songs on the radio from Lubbock or Wichita Falls. The exciting Saturday afternoon football games and the raucous Saturday nights are over, though the headaches may linger. Sunday church is done too, though the radio services will be going on late into the night. It is a peaceful time, as the Lord's Day was meant to be. But on Sunday, October 2, fear of a fall storm was strong in Paducah, Texas. Eighty percent of the area's sparse moisture comes suddenly from thunderstorms, but so does 100 percent of the area's tornados. There were reports of clouds that afternoon, a warning sign that sent a team of volunteer firemen scurrying off into the countryside as storm sentries to watch for those whirling black clouds that make people pray and hide, not necessarily in that order. Being a country fireman used to be a fairly easy job—an occasional fire, a couple of parades each year with sirens wailing, and an annual firemen's picnic with free rides for the kids on the big red truck. But today's rural firemen have more sophisticated equipment, chemicals, and techniques to master, which they practice, with the absentee landlord's permission, on some of the many abandoned farmhouses around town. Blazes have become more numerous and more troublesome, especially wind-whipped grass fires on the overgrown land and fires in the empty houses abandoned by financially pressed farmers.

Fire wasn't the worry that Sunday, though. Wind was. Paducah's radio-equipped sentries had fanned out of town several miles to the west and south looking for trouble in the sky. No one ever thought to watch the north, and Highway 83 running straight down from the Dakotas and Minnesota.

Joyce Hall was having a problem with her hair. She had washed it earlier that afternoon, but it was still wet, and limp. She couldn't go out like that; it was chilly and she'd catch pneumonia. Every time she sat down to do something about the mess on her head, someone would call on the radio. A fire sentry to say all was clear, no funnel clouds over toward Matador or down toward Finney, or a local policeman checked in, or the phone would ring.

Joyce was not well off, not even before the divorce. She and her ten-year-old daughter Lucrezia lived with their cats Felix and Rocky—and "Rocky"'s six kittens. The ten of them lived in the Paducah City Hall at the corner of Backus and Tenth streets. They occupied a tiny, windowless apartment across the marble hall from the drinking fountain and just down the dim passage from the potted plant that was taking a coon's age to die.

Joyce earned her keep there manning the radio for local police and firemen. She'd done it for years. It was a bother sometimes, but she liked being involved, liked being among the first to know things, as police dispatchers often are.

Joyce walked to her apartment door and looked down the hall to check the weather. It was getting dark early. Big drops were starting to fall on the empty pavement outside. A storm was very close. Joyce was not fully dressed, but no one ever came to City Hall on Sunday. She left the door open and went across the room to set her dresser mirror down on the table by the radio microphone. Joyce had her curlers and bobby pins there too. Lucrezia was on the floor near the door, jabbering away at the kitties.

With an ear to the radio and her back to the door, Joyce sat down and concentrated on her hair. She'd have to get dinner soon. She had her arms upraised wrapping her hair around the plastic tubes.

"Mommy," cried her daughter. "Mommy. Mommy. Mommy."

"Mmmmm," said Joyce, her mouth full of bobby pins.

"Mommy! Mommy!"

Then, in the mirror over her right shoulder, Joyce saw him standing there silently watching her back—a six-foot-tall commando. His head was shaved. His arms were bare except for the tattoos. He wore camouflage gear and heavy boots. He looked empty.

Joyce turned in the chair and looked at him, afraid to be afraid for fear of what it might ignite. And wondering, for a hectic instant, how to reach Frank Taylor real fast.

"Is this the police department?" he asked.

"Yes, it is," she said, swallowing. "I'm the dispatcher. Can I help you?"

"I've come to turn myself in," he said.

The flight was ending.

Sheriff Frank Taylor was watching television at home about then. He lives on one of Paducah's paved thoroughfares, Sixteenth Street between Gaber and Birdy, in a one-story brick bungalow with a plaster lawn ornament out front of a petrified Mexican taking an eternal siesta under a giant sombrero. Frank lives there with his wife, Jimmye, and an ancient Chihuahua named Cocoa that Frank has trained to sit up like a bear and beg. Their oldest child, Vince, went elsewhere to seek a living, and thirteen-year-old Jody dreams of leaving.

That afternoon Frank's Dallas Cowboys, America's Team, had come from behind once again and beaten the Vikings, 37–24, way up there in Minnesota. Frank Taylor isn't much for cities like Minneapolis, or even Dallas for that matter. Too crowded. "The only things I want out of them big cities," he says slowly, always slowly, "is me."

Frank Taylor always means what he says but rarely says as much as he means. He prunes his words before they come out. Saves having to take any back. He says a lot in a few words but never says as much as he knows. Even the prairie dogs know this is the best way in the country; anytime you are out in the open, physically or verbally, you are in danger. So keep your eyes peeled for hawks in the sun, your ears alert for snakes in the grass, stay by the trees, keep your mouth shut against unseen ears, and be careful whatever you do. Don't look for unnecessary fights, but never back down when one comes. When attacked, strike back, instantly fearsome. Never give an enemy a second chance. That's what comes with growing up in hard country, one of five sons of a hired hand named Marion Taylor who once earned $1.50 a day in the fields and was glad for it.

Frank quit school in the seventh grade, not because he wanted to but because he had to. The Taylors needed his hired-hand wages. Frank could bring in five dollars a day for just seventeen hours of tractor driving and cotton hoeing. "In order to live," his father told all the boys, "it is necessary to get out and work hard. And things'll take care of themselves."

At eighteen, Frank was driving trucks. Then he was jumping out of airplanes over Korea, shooting people for his government and the United Nations. Over the years he would work for a gas company, and he even tried his hand at farming his own land for a while and working part-time for others. There were thousands of farmers and farm workers there then, but things were changing in Cottle County and elsewhere. Machines were doing a lot of the field work that had required human hands before. Chemicals did the weeding. Everything cost more too, even loans. And folks just couldn't make it on their little old place, even with five sons. Some farms and ranches got bigger. The machines got bigger, and the problems for all the other people got bigger. Time was, on a Saturday night, Paducah's square with the three movies—two English and one Spanish—was jammed with people eating, drinking, singing, cruising, dancing, relaxing. But over the years the population got smaller, steadily, down below twenty-nine hundred souls on the vast landscape. So the stores got smaller and fewer. The

Paducah Post's circulation fell. The jobs dwindled down past 725 in the whole county. "This town isn't going downhill," says Frank Taylor, "it done gone downhill when I was a kid."

Like a lot of folks, Frank had to go away for a while. He worked in the oil fields, with lots of overtime, to pay off all the debts from his farm, which went belly-up; the same thing had happened to his father. Then in 1976 the sheriff didn't run again in Cottle County, and Frank needed a steady job and he knew guns and the folks and the land. He got elected and was responsible for law enforcement across the entire 901 square miles of the rough-hewn county named for an Alamo martyr.

Wearing gray slacks with a big belt buckle, a yoked western shirt with flap pockets, a cowboy hat, and a wooden-handled Colt .45, the trim, sideburned sheriff patrols his dusty domain in a cruiser with a back-seat mini-jail, a minor armory in the trunk, and a teddy bear on the visor. Driving eighty-five miles an hour, the sheriff can steer with his wrists while telling stories to friends along for the ride.

It is a hard job at times and an easy job at others. He gets $1,250 a month for being on call around the clock, no vacation for years, just an afternoon or morning off here and there when things get slow. Most of his calls concern property crimes, thefts, break-ins, even some cattle rustling, though that's backed off with the fall in meat prices. "I got the same crimes out here as the police got in the cities," he says, " 'ceptin' a whole lot less of it."

Killing used to be a regular Saturday night business at the bars or the pool hall. Most of those establishments are gone, but there are still some fights around, especially among Paducah's Hispanic population, who used to be called Mexicans. Whatever their name, the sheriff knows those families carry grudges for generations. "They don't get over anything real quick," he says. Fact is, there was a Mexican wedding not long ago, over at the VFW hall. Big wedding crowd. A Mendoza was getting married. There was some drinking. And then some fighting. And then a member of the Makata family shows up with a .38. And he starts shooting all over. And he hits a few folks. And the sheriff gets called. Makata opens up on him. "Well, I tell you," the sheriff recollects softly, "you get people shooting at you, they get your attention right quick."

What happened then? "Well," says the sheriff, sucking on his toothpick thoughtfully, "I shot him." Señor Makata got over his grudge real quick.

Sometimes though, being Cottle County's sheriff is quiet work. Plenty of time to catch up on paperwork, do some driving and show the badge around, watch for city folk passing through, make sure they keep right on passing through if they got mischief on their minds. Out-of-towners are easy to spot, and not just 'cause of their clothes. Out-of-towners are the ones who lock their cars. "Watch your step here now," Sheriff Taylor warns one out-of-towner at dinner time, "we're gonna roll up these sidewalks here in just a minute." Then he puts the toothpick back in his mouth and smiles, slightly.

Being a sheriff can be a tricky business in a small town, wielding authority when necessary, but no more. A sheriff's got to make sure the things that are important get done and the things that aren't don't get in the way. That requires a lot of diplomacy. Arresting a drunk in a rural county isn't that simple. For that drunk is also likely to be a friend, a son of a friend, a husband of a friend, or all three. He's a voter, too.

Being sheriff also requires careful socializing. Frank might sit around his tiny office in the county courthouse and chat about fishing with some rancher who wonders if it's true that J.W. is selling that piece of land or when the next Rodeo Committee meeting is set. Or Jewell Gibbs, the justice of the peace and Wesley's wife, might wander back from her nearby desk if there were no traffic citations to handle right then. Jewell was the one who used to have a pet prairie dog. She'd say, "Sit, Jackie." And the little critter would sit up on her table and chatter away. Jewell had to give him away though. He had a habit of chewing things, which was okay and natural and all until he took to chewing up Wes's fancy, hand-tooled leather belts. Jackie was lucky to escape that day with his little furry head still on.

Often, Sheriff Taylor can be found at Burns Kountry Kitchen, out there on the highway, where the coffee is free for police and the jam comes right out of the clear-glass jar it was made in last fall back in the warm kitchen. The Paducah news hasn't been so good around those tables in recent times, mostly about folks in trouble or leaving. But news was about to break in Paducah, news about someone in trouble who came into town and for a few flickering moments put Paducah on the map, or at least on the television screen, which sometimes seems like the same thing these days.

Ronald Kay Simpkins, a part-time police officer, was the first to arrive at City Hall that Sunday evening. An excited Joyce Hall had called anyone in authority to get down there. At first, given his appearance, she'd thought Steve Jenkins was an AWOL soldier from some military base. He started to cry pretty quick in her apartment and she got him a fistful of tissues. She was comforting him and telling him everything would be all right, but right away he had said something about accomplice to murder. Up in Minnesota or somewhere. Joyce forgot about her wet hair for a long while.

Kay had frisked the boy while the rain came down hard outside. Then Randell Bockleman arrived. Randell had been a farmer for years but had seen the financial writing on the barn wall three years earlier, and he'd gone out and found himself a new, safer line of work: town policeman.

When Frank arrived, the three men began to question the teen-ager. Around Joyce he'd been bawling like a frightened little boy, but with the men Steve straightened up. He was cool, real cool. No sign of fear, just calm, short answers. There's been a couple of murders, bankers, up in Minnesota, he said. He'd been behind the garage when it happened. Heard a bunch of shots, someone yelled, "He's got a gun," and then more shots. Steve said that when he came out, his father was walking back around the house from the front yard and he said, "I fixed that son of a bitch Blythe."

This was all confusing horseshit to the three men. They had never heard of Ruthton or this Rudy Bligh guy, although it rang a bell for Joyce, who called her mother to check the newspapers. The police didn't know about any Minnesota murders; they hadn't read that morning's *Avalanche-Journal* from Lubbock so they didn't know what to fish for in the kid's statement. The farmer-banker problem seemed credible enough. The kid said the banker had taken his dad's land away even though he'd made the payments on time. Sounded familiar.

"And speaking of your father," the sheriff said, "where is he, boy?"

It's still known as the old Goodwin house, even though no one, including Ernest Goodwin, has been able to live in the place for nearly twenty years. It's a little over three miles north of Paducah on Highway 83 and six tenths of a mile east on a nameless gravel road that runs straight out of J.R. Bratton's front yard. The old farmhouse burned down long ago; the volunteer firemen couldn't get there quick enough. Goodwin had been in a wheelchair from a car accident in the mid-1950s, so he couldn't fight the fire. Ernest had moved to town, and around 1975, the Brattons, J.R. and his brother J. Wayne, the dentist, bought the land for grazing their breeding cows. Nothing was left of the house but its cement slab foundation, now hidden behind a couple of cedars going wild from their civilized days of decorating a front door.

Pretty near everyone thereabouts knew where the old Goodwin place was, everyone except a couple of strangers who arrived after dark on Saturday night. They were specialists in abandoned houses; in fact, one of the two men in that white pickup truck with the rusty South Dakota license plates still felt he owned an abandoned place up in Minnesota, and the two of them had been staying in such places for the last couple of days on their way south.

The two men didn't go straight to the Goodwin

house. Early Sunday morning, their dirty truck, heading south, drove right by J.R.'s. The older man in the truck said there was no destination, they were simply gonna go away, back to the Sunbelt where things seemed successful.

At first, Jim Jenkins thought he was running away from that bloody farmyard and the police and bank people. Jim was headed to no particular place. South, just south. Maybe he would know it when they got there, a safe, sunny place where things would work out right for once. It never occurred to Jim that the appointment was already set, and he was right on time.

Most out-of-state trucks blast right on through Paducah, especially at night when everything looks abandoned. They pretty much stop at Garrett Street, which becomes just a red blinker by midnight, and then they grind back up through all their gears on their all-night way to somewhere else. But this white truck turned off 83 in downtown Paducah and cruised the square and the dark streets for a bit, looking for gas or something. If Sheriff Taylor or Randell had spotted the South Dakota truck, he would have kept a close eye on it—trouble looking for somewhere to land. Anyone up at that hour in Paducah is up to no good. But no one remembered seeing the Chevrolet or paying any attention as it circled the square past the courthouse, the abandoned gas stations, the fading lawn tributes to the girls' track team, and the florist shop where the material for a Jenkins funeral wreath was sitting, waiting.

The father and son turned back north for three miles or so, and at J.R. Bratton's front yard, they turned east. They parked by a cow pasture, or what passes for one in the arid parts of Texas, at the open end of a dusty, dead-end lane. From that little rise, a field-sized knoll really, anyone can see the landscape's spread. It was a good place to sleep. There, with field rats noiselessly scurrying around beneath them, leaving meandering trails in the sand by the truck's wheels, the two men dozed in the silence for a few hours. The pink of Sunday's first light in the east uncovered a building of some kind about two hundred yards up the sandy drive. It looked like an abandoned garage. Probably the house burned down. They backed the truck in behind the garage by the old tree, snapping off a large, low dead limb in the process. They took some gear and their scant food over to the cement slab and sat down behind the cedar trees, with a view of the surrounding county that could not see them.

They had heard the news of the killings and the manhunt on the radio during their long droning nights on the road. Many of the newscasts carried descriptions of them and their truck, and sometimes even their old license plate number. Dot by dot, their pictures were flowing electronically into every newsroom in the country, thanks to old Abe Thompson. No matter how far south they went, neither Jenkins could tell anymore when a smiling gas station attendant was just trying to be friendly or when he was stalling until the state police quietly arrived out front and, guns drawn, firmly asked to see a driver's license.

Steve says he remembers his father growing more tired and irritable as the day wore on and the big country seemed to grow smaller around them. Steve says he suggested they turn themselves in, and he remembers his father saying he would never go to no jail, and if he was going to die in a blazing gun battle, he wanted to take some others with him. Steve says his father grew even angrier when the son announced he was going to surrender—there was no point in going on. Steve says they argued for a long while that afternoon. Then they sat down on the cement slab for their last supper, half a can of cold beans each and a few swigs of warm Pepsi from a plastic, two-liter bottle they had bought two lifetimes ago in Pipestone.

Late that afternoon, as Rudy Blythe's friends and relatives were gathering back in Minnesota for a memorial service, Steve stood up to leave the cement slab. The sky was darkening in the west, where the future seems to come from these days. Steve says his father had often threatened suicide since the divorce, saying there was not much to live for anymore, and sometimes he threatened to take his son with him. But this time, Steve said he was determined to leave. Somehow he managed to get the pickup truck stuck in the sand. Jim Jenkins helped his son out, but before the boy drove off, his father took the shotgun from the truck. For a man who planned a shoot-out until death, Jim didn't take much ammunition, and when Steve looked back across the flat land, his father looked very small.

Not long after, the storm arrived, starting as big lazy drops and then turning into sheer sheets that waved in the wind like the folds of a watery curtain suspended from the sky. Jim Jenkins sat out in the downpour, his denim clothes growing dark and heavy. That was the least of his concerns; Jim didn't have enough time left to catch pneumonia. His old farm was gone; all his farms were gone. His wife was gone. So was his daughter. His big Texas savings were gone. His new farm dream was gone too, shattered into a thousand pieces like a windshield when something hard hits it. Jim's eyesight was going, too. The future of the forty-six-year-old man had died in that farmyard right along with those bankers. Now even Steve, his son, his farming partner, the one person who kept coming back to him no matter what, now Steve was gone in his pickup.

Jim picked up the old shotgun and the new flashlight, and, aiming his tunnel vision down at the ground in the gathering twilight, shuffled through the sand from the safety of the bushes to the dangerous open area toward the end of the road.

The heavy rain had stopped. The Texas soil smelled good and moist, not as good as the thick black gumbo back home, but still fresh and clean, recharged to fuel new growth. Jim could feel the ground rising slightly beneath his feet as he approached the junction of the lane and the dirt road. After spending four days in a cross-country flight avoiding people as much as possible, Jim Jenkins was insuring that he would be found, finally. Jenkins walked up the little knoll where he could see, or sense, the surrounding countryside. He turned the flashlight off.

Jim carefully removed his glasses then and folded them up. It was a simple step and a natural one for eyeglass wearers about to go to sleep. But it was a step that would later come to haunt his son and his son's legal defense.

Jenkins stepped back a little and turned toward the west. It was brighter there. That's where the wind comes from. That's where the sun sets on opportunity. Milking time back home. With his left hand Jenkins moved the broad muzzle of the twelve-gauge shotgun up toward his mouth. The barrel was cold and hard, very hard, as he slid it between his lips and teeth. It tasted of oily steel. His right thumb moved to the shiny trigger, which was cold as he pushed it ever so gently. Darkness was falling.

No human ears heard the thunderous blast as it rolled out in all directions over the massive countryside. A grazing cow or two raised its head a little to take note, then the natural cycle continued.

Jewell Gibbs loves that countryside. It's so tranquil. And when life as a wife or justice of the peace or Cottle County coroner gets too hectic, she often drives out into it and just sits in the pickup truck for a long while, soaking up the peacefulness. Dusk is her favorite time, when the coyotes are yowling to each other, a pack wailing and yipping on one side and then another answering back on the other side, and both getting closer to mark their turf. One time out at the Masterson Ranch, the coyotes were yowling all around and a colony of ground squirrels was frolicking on one side of the road while a little calf, minutes old, was wobbling about under the nurturing nudges of his mother's muzzle. "It was plain wondrous," she recalls.

Not long after she got back home from the ranch, her phone rang. Jewell raced a few miles out of town to a clump of trees on another dirt road. Sheriff Taylor was standing in the sand looking down on Mickey, his brother, lying there in a circle of brownish-red dirt grown darker by leaking blood. A deer rifle lay not far from the large hole in the man's left chest. Mickey had been depressed about a number of things, including his teen-age son's crippling automobile accident the previous week. They told Frank Taylor his forty-six-year-old brother had died instantly that spring night.

Now it was fall, and Sheriff Taylor was asking another teen-age boy about another forty-six-year-old man. "You sure it was this road?" said the sheriff, slowly turning to the east on the gravel road by J.R. Bratton's house. From the kid's description, it sounded like the old Goodwin place.

The kid in the back mumbled yes. Steve Jenkins was handcuffed, behind his back, just in case. Kay Simpkins was sitting next to him. Randell, the town cop, was in the front seat with Frank, who was thinking. This bald little fucker in the back seat was awful cool and calm for someone claiming to be a murder accomplice. He'd been talking about how his father was threatening suicide, sure enough. But that didn't ring right to Frank. Why would a boy leave his troubled daddy alone with a shotgun? Most sons would have stayed, or taken the gun away, Frank figured.

Maybe this was all bullshit, another ambush setup right here on Frank's own turf. Maybe the kid was a runaway soldier or some kind of weirdo doper, a hitchhiker who got a ride from some middle-aged man and then wasted him for money and was trying to make it look like suicide, and maybe would take out a couple of cops on the way. Maybe there were more weirdos out here. Frank Taylor, the ex-paratrooper, slowed his car to a halt and looked around for a moment at the kid, at Kay, and then at Randell.

"Shee-yit!" said the sheriff as he got out of the car and opened its trunk. He picked up his machine gun and returned to the front seat. He put the cold weapon in his lap. Then he slowly drove toward the Goodwin house.

Frank was looking hardest, so he saw them first, a pair of booted feet sticking out in the road, very still. It was strange, almost comical. The man's right heel was propped atop his left toe. What Frank saw next was not comical. Most of the guy's head was gone. Pieces of skull and brain were scattered about. The sand was dark red running from the neck in large splotches, downhill.

Frank saw the shotgun. He held on to his own gun and looked around anyway, very carefully. There was just one fresh set of footprints in the dirt, so the victim had come here alone after the rain. His clothes were soaking wet; he obviously had sat outside through the whole storm. The tracks came down from the old homestead, walking up on that knoll and facing west. Wonder what he was looking for out here in the open?

Then a few footprints back down the knoll, where the man stood a minute. The blast of the gun—Jesus, a twelve-gauge right in the face—had blown the man around toward the south, and he'd fallen back on the ground, stiff as a board, a scarecrow in denims, still a little warm. Whaddya suppose he'd want a flashlight for? It wasn't dark yet. And look, the glasses.

"Hey, boy, did your daddy wear glasses all the time?"

Yeah, well, he won't anymore. The poor bastard musta known what was coming; he put the glasses in his pocket. Frank could see only one eyeball, hanging out.

When J.R. Bratton drove up, Frank was looking at the dead stranger in the dirt, seeing someone else somewhere else, someone much more familiar. J.R. was puzzled by the small crowd on his property. The place had been quiet and empty just thirty minutes before when he drove by. J.R. saw the body and figured some excited dove hunter had accidentally blasted himself. Then he saw the handcuffed kid casually looking out of Frank's car. He knew that kid from somewhere.

Frank didn't introduce them. He was busy checking out the area with Randell and radioing for the coroner and the Texas Rangers. No doubt old Leo would be arriving soon.

"J.R.," said Frank, "it looks like you got some visitors on your place. This kid says he and his daddy were involved in a couple murders up north and—"

"That's it!" said J.R., who had seen two family photos and read an interesting story in that morning's newspaper about a wanted farmer lashing out at a couple of bankers. It was the kind of story every rural resident would read, filling in his own feelings and paragraphs at every turn. "That's the kid that's wanted," said J.R.

Abe Thompson was at home in Minnesota when the call came. He had to look at a map to find Paducah—from the middle of nowhere in Ivanhoe, Minnesota, to the middle of nowhere in Paducah, Texas. When the two rural sheriffs talked, Steve Jenkins was sitting on a hard wooden chair in the Texas county's courthouse. He was making a voluntary statement, and Frank was getting it all down with Leo Hickman. Laconic Leo is a tall fellow whose short white hair, parted in the middle, makes him look like somebody's uncle from the 1930s. Leo is an employee of the Texas Department of Public Safety, one of the fabled ninety-six-member detective force better known as the Texas Rangers. Cottle County is part of his territory. And not much there escapes his notice, even though Leo's silver-rimmed eyeglasses help disguise the fact that he was shot in the eye during a forgotten gunfight years ago.

After they'd read Steve his rights and started the interrogation late that evening, the kid said he was real hungry. Randell went out for food, and Steve wolfed down a big cheeseburger, a pack of french fries, and a vanilla milkshake, which struck Sheriff Taylor as kind of strange since he couldn't eat for days after his own father died, and the old man had passed away in a hospital, not swallowed the wrong end of a belching twelve-gauge.

Leo and Frank found the Jenkins truck parked near City Hall in the little lot behind the dentist's office where J.R.'s brother worked drilling and buffing and pulling teeth. They found a small armory in the truck, including several guns, a steel helmet, ammunition, some kung-fu throwing stars, and a .30-caliber carbine that would attract the interest of Minnesota authorities. Looking at all those killing tools, some of the police were thinking how ironic it was to have this eighteen-year-old kid who was too young to buy a can of beer legally in Texas but plenty old enough to walk around with a deadly arsenal like this.

The truck and all the weapons and Steve were taken

into Minnesota custody after Abe Thompson arrived the next day in a farmer friend's chartered plane. With him were two experienced BCA agents, Bob Berg and Mike O'Gorman, who would follow the Jenkins case to its end. It was Mike, a savvy veteran of law enforcement battles and a heart bypass operation, who, after that long, low-level flight from the sprawling flatlands of Minnesota to their broad-bottomed counterparts in Texas, thought he might rather face the entire Ma Barker gang alone in an abandoned farmhouse than go back into the stormy autumn skies over the prairies in a small private plane.

Right behind that private plane, landing at Paducah's little airport just outside town, came other chartered aircraft and then rented cars down the highway as the television crews descended, a crowd of licensed curiosity-seekers in a big hurry with many questions and few manners. That Monday and Tuesday Paducah was like many of the country's obscure and sometimes isolated communities that are suddenly thrust into the national spotlight for a moment by an event over which they have no control and probably even less understanding. Some native son becomes an instantly notorious assassin and strangers with cameras and notebooks appear from nowhere looking for relatives, pals, and high school yearbooks. A vicious storm strikes and the next morning some man in a pressed safari jacket with a microphone and skin that is too tanned for this time of year wants to know exactly what it was like last night and how it feels to be showing the world today's pile of worthless rubble instead of yesterday's handbuilt home. When a woman starts to cry nearby, the camera turns immediately, tears and yelling being very valuable professional currency back in the editing room. "Thank you very much," these visitors sometimes say, as if the destruction was laid out just for their arrival.

At first, the local people kind of like the TV attention. It's flattering, different, and a little exciting to be converted into a movie and bounced off satellites to fill up the air space between speeding new cars and talking elves who make cookies. It's flattering to have someone important—he must be important, he's on TV every night—actually seeming to pay attention to their thoughts and concerns and hopes and dreams, even though this important someone can't remember their names an hour later, he's asked so many people the same questions.

According to rural etiquette, no one can appear to seek or like all this attention, although it is socially acceptable to rush right home and watch the news, privately, to see if one's existence was confirmed by an editor looking for thirty-four seconds of drama or quick quotes on tape to lead into the sports report.

The local folks could understand the farming hardship part of this suicide story sure enough, the bankruptcies and divorces, the heart attacks and lone gunshots out by the barn. They understood the running financial feuds between bankers and farmers. Everybody knows a good banker story: "You know what the difference is between a dead skunk on the road and a dead banker on the road? There's skid marks by the skunk." Sometimes the Texas farmers had some questions of their own for the reporters, such as: How could this Jenkins guy be as bad as they say when he had no money and no gas and all those guns and didn't rob nobody here to get some? Maybe he did kill somebody and that's wrong, no doubt. It's especially wrong in America if you kill a banker, if you know what I mean. But he coulda killed others here, surely robbed 'em, and he didn't. That don't sound like no hardened criminal to me.

On Monday the trouble was that no one in Paducah knew anything about Jim Jenkins or Steve for the scrambling TV stars to interview them about. People from far away wanting to talk about two men from far away in a place that never heard of neither. The reporters from Texas didn't know too much about the Minnesota happenings and the Minnesota reporters and crews didn't know much about the Texas angle, and Abe and Bob and Mike and Leo sure weren't talking in front of any cameras about what they knew. So instead there was a lot of scenic footage on the news that night, shots of the stained suicide scene, the

Goodwin garage, the City Hall and downtown, maybe some cattle grazing in a very big field, a TV correspondent with hair mussed by all that wind standing by the county courthouse or the white pickup truck, and a clip of Cottle County Sheriff Frank Taylor outside his office saying, no, the kid didn't seem surprised when he saw his father dead in the road. But the sheriff firmly believed Mr. Jenkins's death was a suicide anyway, unless Batman sneaked up on him in a helicopter.

Sheriff Thompson and the two BCA agents questioned Steve near Frank's office for more than two hours on videotape. They went over all the events of recent weeks. They searched the truck with great care, inventoried the arsenal, and began to form the legal and criminological hypotheses that would become the foundation of the circumstantial case against Steve. It was not going to be an easy case. Everyone who knew what happened at the farmhouse that Thursday was dead, except the defendant. He had marked his eighteenth birthday on August 21, five weeks before the shootings, so he was an adult in the eyes of Minnesota law. He would be charged with a most serious adult offense: murder.

They took Steve home in Abe's friend's little plane. It was Steve Jenkins's first airplane ride, and he sat there, in handcuffs, his excited face glued to the window and the view of the big land where he had so recently been hiding. Mike O'Gorman volunteered to give up his seat on the private plane that day. He rode back to Minnesota in one of the TV crews' larger chartered planes.

Bob Berg and Mike O'Gorman, and for a while a few others, began the reconstruction of the Jenkins men's lives and movements and Susan Blythe's recollections and exactly what might have happened in those few fleeting seconds on that rainy morning in that farmyard near Ruthton. This was a grisly process, reenacting two deaths, firing guns at windshields, and reliving, emotionlessly, the moment of death to pinpoint the sequence of events and the killer's identity.

It is an all-consuming task, this conscientious reassembly of the past. The agents see that there are victims in everything, most of them unsung, painfully caught up in the crimes and their aftermaths. People like Jim Jenkins's parents, who worked hard all their lives, covered their son's bad debts, and who one day woke up and found they had an emotional choice: they could turn on their only grandson for the sake of their dead son or they could forsake their only child as a killer and support their grandson.

The agents isolate incidents from one long, complex life and with them paint a portrait with words on paper for the justice bureaucracy; this, they must conclude, is what that entire life was about. The details are true. They are also incomplete, and the agents wonder at times what their own lives would look like if compressed like that in a file for strangers. The subject's life also comes to dominate the searcher's life so that even on infrequent days off, the detective's mind is still pondering, still wondering.

Jim Jenkins, a dead man the investigators would never know and have seen literally in pieces, becomes an intimate, controlling fixture in their daily lives and the daily lives of their families. They come to talk of him as they would a friend, by his first name, and often in the present tense. They find themselves inside his personality and actions, stumbling upon some of the same obstacles he encountered, and then trying to imagine from their notes and photographs and their professional intuitions which paths this troubled friend-for-a-moment would see and which he might take and where they might look for more information.

It all can cause agents to be inattentive to the details of their own personal lives. And there is a great deal of grinding routine and professional challenge in this work, sorting out the blind alleys from the promising lanes of inquiries, the leads from the misleads. After one particularly demanding case, there can be a kind of postpartum depression as the agent returns, however briefly, to an office routine or to investigation of less challenging cases.

The investigations can touch deep feelings within the investigators, forcing them to confront emotions they never knew they had or chose to ignore. The investigations can involve relationships between fathers and sons, ringing unconscious bells about the loves and sometimes simmering hatreds that link those adjacent generations of males. The detective can sometimes find himself investigating himself and his own feelings and motives right along with a criminal subject, an intense personal pain paralleling the professional problem, but one that goes unchronicled in the criminal files.

This discomfort can be and usually is covered, however, by denial, flippancy, and black humor during the long, lonely evenings of free time on the road in the antiseptic motel rooms (first floor please, and even numbers only, for good luck). There, the potato chip bags accumulate and the TV is left on, soundlessly, for company while the day is digested, the new day planned, and the wife telephoned.

Sometimes, once in a great while, all of this time and thought and work come together, and a prosecutor and a jury and a judge see the same things the same way and reach the same conclusion. There is quiet satisfaction in feeling, briefly, that justice was done before rushing on to the next case.

There can be great frustration in all this work, times when everything points to guilt, but the prosecutor, perhaps preoccupied with personal problems, can't explain it powerfully enough or the jury can't see it or the judge won't agree or the detective screwed up somehow. That is the way the imperfect system works, imperfect but better than any alternative, they tell themselves. Those disappointments, which are legal triumphs just a few feet away, pass with some time and detailed immersion in the next case.

The unsolved cases don't pass so quickly. In them, everything is clear and obvious, but the crucial link is absent. A missing wife. A strange husband. A little blood somewhere. A lot of suspicion. Perhaps a first wife who died under circumstances that seemed natural then. A lot of suspicions. But no body. Sometimes one of these cases can become a personal crusade for an agent, the kind of case perhaps where the missing woman reminds him of a mother and pricks his sense of injustice too much not to come back to it time after time, pulling the bulging, memorized file out once again whenever there's a free day or a weekend or even an empty morning, there being no statute of limitations on murder. These perplexing riddles are called career cases because they will take a full career to solve, if they ever are.

In one sense, the Jenkins case was an easy one. The bodies were there. The gun was there. A witness placed both men at the scene. But in another sense, the Jenkins case was very difficult. The evidence was pretty much all circumstantial. The only living survivor of the four, the only one who really knew what happened, said he didn't know what happened. He said he was hiding behind the garage the whole time and emerged only after two bursts of gunfire to see his father returning from the front yard saying, "I fixed Rudy."

Berg and O'Gorman had already gone over and over the crime scene in Ruthton. They had gone over the truck and Steve's few possessions, including all the weapons and a note they found for Darlene Jenkins in the youth's wallet, apparently put there just in case something happened and Steve wasn't alive to talk. "I love you Mom," it said. "Tell gradma [sic] and grandpa and Mickey that I love them. I'm sorry that this happened. Please forgive me and daddy I love all of you for being there and helping when I needed it. I wish things could be different If I could change what happened I would I wish I could be with you. I love you. Steve." The agents, along with Sheriff Thompson, also went over and over the whole situation with Steve during that long questioning on videotape in Paducah. He mumbled and kept his head down. But he was cool and assured. He also sniffled after each answer, which might mean something or might not.

The agents ran over plenty of easy stuff with him at first, all the legalities about voluntary questioning and understanding what he was signing. They got into his date of birth and address and phone number and his childhood, much of which they already knew but which helped later to judge Steve's credibility. They hoped that as he respun the strands of the web they knew, he might leap into a new area and fill in a gap or two. Steve talked about their problems in farming and working with his father in Texas and how he loved military things. All the clothes he had were camouflage and khaki green. He said he took his carbine everywhere, even to bed at night. He talked about not getting into the marines because of his missing spleen. Steve explained why he'd quit high school halfway through his junior year; they had changed his shop teacher and he didn't think he was getting much out of it anymore.

From different angles the questioners gently but persistently came back to the morning of the murders and the days leading up to it. The only time Steve wavered was on the subject of his father and his poor health. He had diabetes, you know. And night blindness. And what both father and son had done together, living and working together and all, and his father sometimes saying he didn't have much to live for anymore, not since his wife left and the banker took away the farm and now was telling everybody else around not to loan Jenkins any money. Steve got a little weepy then.

The agents made a note to check the eyeglass prescription. As the questioning progressed, they boosted the pressure a bit. They said, well, uh-huh, Steve, that was interesting and all. They'd like to believe it, they really would. But they had some problems with that story. It sounded good, but first of all, Steve, it didn't matter if you were there and pulled the trigger or if you were there and didn't pull the trigger. You were there. You were an accomplice. And in the eyes of the law when a forty-six-year-old and an eighteen-year-old walk into a situation like that, with all those guns and anger in that isolated place with two defenseless people who become dead as a result, then both that forty-six-year-old and that eighteen-year-old are equally guilty in the eyes of the law. Plain and simple.

Steve said he shot no one. That may very well be, they said, but it would be a whole lot easier on him if he just said what really happened, you know, just came clean and got it over with. Steve stuck to his story. They had rented the land and house with the Texas money. Got some machinery, too. Then they had gone from bank to bank and cattle dealer to cattle dealer throughout September looking for money and cows. Time after time they had been turned down for credit. Sometimes Steve was there when the man said no. Sometimes he was in the truck when his father returned and said the man said no. Steve said it became a daily occurrence. He said his father was real discouraged and he blamed that son of a bitch Rudy Blythe for everything. Steve said he couldn't understand how bankers could get away with taking things that don't belong to them when other folks would get arrested for such stealing. Steve had heard that Rudy had taken away fourteen other farms. And then one day, the day before the shootings, Steve said his father came home from the last credit turndown and announced he had used a fake name to make an appointment with Blythe out at the old farmhouse the next morning. They were going to rob Rudy, rob him and scare the hell out of him, Steve remembers his father saying.

Uh-huh, the agents said, fine, okay. But that really didn't fit with some of the things they had started to learn from other people. Steve said, well, he didn't know about that, but he was scared as hell out there at their old farmhouse and he knew he didn't shoot nobody. The agents said they had a different idea of what happened. They said they thought Jim and Steve Jenkins made a deal before the suicide. They thought that Jim had maybe made Steve promise that no matter what happened, no matter what anyone said, Steve should always maintain that his father did the shooting. No one else was there. How could anyone prove otherwise? With his father about to be dead, Steve would be free. The agents and Sheriff Thompson said that would explain a lot of things, and that's how it looked to them. And Steve owed it to his father's name to let him rest in peace in his grave, to come clean with the truth.

Steve was crying now. He said he hadn't wanted to leave his father. The agents said they understood.

They warned him about his conscience. Over the years, they said, "it'll eat you alive."

Later, reviewing the tape and watching Steve's reactions, the agents came to believe he almost broke then, almost said something that would have made life a lot easier for a lot of people in the coming months. There was a pause. But Steve stood steadfast. He didn't know about any other theories, he said. When he had told his father he wanted to surrender, his father had said, Never—he wouldn't go to jail. They had argued, Steve said, and slept, and then Jim Jenkins had said his boy could go. He wouldn't stop him. But if Steve left, Jim said, he was going to kill himself. There wasn't much else to do. He took the shotgun from the truck because, he said, it would do "the best job." Steve only knew what happened out there, he said. He hadn't killed anybody ever. What did these guys want? Did they want him to say something had happened when it didn't really?

"Well, fellas," said Thomas L. Fabel, "I'd say you have your work cut out." The tall, deep-voiced Fabel unfolded out of the chair where he'd been watching the Steve Jenkins videotape with O'Gorman and Berg. He was deputy attorney general of the State of Minnesota. He was also concerned. Fabel had learned during his legal career never to be too certain of anything, especially in the fields of law and human behavior. On this day he had more doubts about the accused's guilt in this Jenkins case than he felt comfortable with as a professional prosecutor. Fabel would be the government's point man on this highly visible case, which had even grabbed page one attention in *The New York Times*. Within hours of the ambush, Mike Cable, the local part-time prosecutor in Lincoln County, had asked for help, and Fabel himself was going to need a lot of it if they were to prove the charges against Steve Jenkins. The police had done their part in producing a defendant. Now the ponderous machinery of justice had been set in motion.

Things had quieted down in Paducah, the media's interest having evaporated within forty-eight hours. The broadcasts were forgotten soon after, amid the bigger splashes of more murder in Ireland and Lebanon. The Texas town got back to its slow decay, except for a little incident right around New Year's.

One bleak day two men showed up in a private plane at the community's little airstrip. They introduced themselves to Randell as Louis Taveirne and James Dwire from Minnesota, friends of the Jenkins family, they said. They very much wanted to see the suicide scene, they said, because they were convinced Jimmy Lee Jenkins had left a suicide note that would clear his son of the Blythe shootings. It being nearly dark, the trip to the old Goodwin place was postponed until morning. But when Sheriff Taylor arrived for his 9:00 A.M. appointment, the pair had left word they went on ahead.

Frank was a little worried about J.R.'s reaction should some more strangers, and northerners at that, be found strolling on his land. In addition, since the suicide, J.R. had put in some electric fence right by the old garage. Its steel strands ran from an old fuse box on a nearby pole and poured out enough current, as they say in those parts, to straighten up a part of the male anatomy right quick. When the sheriff arrived, J.R. was already there. He hadn't said much to the strangers except "Howdy" and "Whatchall doin'?" But his eyes silently spoke deep suspicions to his sheriff friend. For people who'd never been there, these two fellows knew an awful lot about the layout. Frank and J.R. were chatting for a moment when Dwire and Taveirne suddenly shouted from over by the old fuse box. "Hey, sheriff!" they said. "Look here! Look what we found." And they came running over with a little piece of pink paper.

J.R. and the sheriff looked at each other silently again. They had both looked in that box several times since the suicide, even though it was a good two hundred yards from the body. In fact, the day after the suicide Frank had scraped a lemon-sized hornet's nest from the leaky metal box with his pocket knife. Now, ten weeks and several thunderstorms later, these two Yankees just happened into town and just happened to find Jim Jenkins's official suicide note on the unstained back of a TV repair bill an eighth of a mile from where Jim Jenkins had blown his head off. These visitors also just happened to have a camera along which they used to photograph Louis Taveirne handing the note to Sheriff Frank Taylor, whose look at that moment would have displeased any Chamber of Commerce hospitality chairman. The penciled note said: "I killed Rudy Blythe the SOB. Steve leaving. Won't listen anymore. A guy just as well be dead." It was signed by James L. Jenkins.

One newspaper headline of the discovery read: SUICIDE NOTE TURNS UP IN JENKINS MURDER CASE; SHERIFF HOLDS DOUBTS. What Sheriff Taylor was really holding was his temper. The two Minnesota men were, as Frank put it, "standing there pissin' on my leg tryin' to tell me it's rainin'."

The sheriff told the men he didn't want their phony note and he suggested they remove themselves, fast, from this private property. "But, Sheriff," said Louis Taveirne, "what should I do with this note?" Frank Taylor paused only a moment before he told the stranger precisely where to put it.

Eight days after Jim Jenkins took his last steps in the Bratton's field, he was put in the ground. The funeral in Paducah was ignored by most of the town and all of the world. Faced with a $150,000 bond for Steve, the Jenkins family told Pat Seigler on the phone that they wouldn't be bringing Jim Jenkins's body back to Minnesota, not right away anyway. Maybe next year. And Pat got to thinking. As the lone mortician in Cottle County, he does a pretty fair business burying folks, between accidents, suicides, old age, and a growing combination of the last two. There'd been four county suicides in the past year, counting Wylie Boyle, the old-timer who shuffled into the hospital room of his bedridden wife and ended her Alzheimer's agony before pointing the .38 at his own heart. Few of Pat's burials carry the prospect of being disinterred, by Pat, a year after they entered the sand. Pat knows about bodies and the sand and the durability of those fiberboard pauper's coffins covered with cloth. So this time the mortician prevailed upon the

county to spend a little more on an indigent burial.
Jim Jenkins got a steel coffin.

The Reverend Emmett Autrey drove the long white
hearse down the dusty little path that separates the
neat rows of mounds out in Paducah's Garden of
Memories Cemetery. There are more than three thousand people buried there, more folks than remain
living in the entire county. Three of the sheriff's broth-
ers are there—the suicide, the car accident victim, and
his oldest brother, the one who burned in the truck.

It was late Monday morning. Nine people had gathered at the hole, freshly dug by Jimmy Branson. No
one who ever knew Jim Jenkins was there. No one
who ever knew him would have recognized his re-
mains. "But God knew him better than anyone," said
the Reverend Autrey of the Missionary Baptist Church.
"It is not our place to judge his thoughts, his emotions, his actions," said the minister, squinting against
the bright sun and holding the pages of his Bible open
against the wind. "We are here simply out of respect
for another human being." Sheriff Taylor was there
and Pat and the cemetery's caretaker, Robert McGuire,
and two state troopers, Leroy Bernal and Randy Rister.
Off to the side stood Thelma Martin and her daughter,
Joyce Hall, the police dispatcher who first encountered the Jenkinses and the only one to send any
funeral decorations, white plastic flowers.

Mere mortals, the minister said, face two final
certainties—to die and to face judgment in the hereaf-
ter. "But that judgment," he said, "brings into it
every secret thing." There was a moment of prayer
before Jimmy Branson eased his growling backhoe up
to the site, next to the final resting places of James
Harbison and W. James Burns, and began pulling the
soft dirt back into the hole. The little stone would
read: James Lee Jenkins, Feb. 4, 1937—Oct. 2, 1983.

"Who can say what this man was really like?" Pat
Seigler was saying. "If they did what they were accused of, of course, it was criminal. But they didn't act
like hardened criminals. Possibly their only differences
were with those who had taken the farm away."

Eleven days later they dug up Jim Jenkins and hauled him two hundred miles to a lab in Dallas for a day before burying him in the Paducah sand again. Something about his eye.

7

THROUGH
THE LOOKING GLASS

"All rise."

Six dozen people stood for the entrance of a gray-haired gentleman in a black robe who would preside over the next few weeks in their lives and the changes they could not know were coming.

The trial of Steven Todd Jenkins began on April 10, 1984, in the freshly painted courtroom on the third floor of the Lincoln County Courthouse in Ivanhoe, Minnesota, 475 miles northwest of Chicago and 15 miles northwest of the murder scene as the hawks fly. It was an overcast, drizzling, cold Tuesday, not unlike That Day back in September. Fog covered many fields and chilled drops of water dripped steadily off the bare branches of the trees, christening the old soil with the first sign of the new spring. That hopeful season was probably somewhere down in Kentucky now, slowly but steadily moving north every day.

For now, in Ivanhoe it was more the end of winter than the beginning of spring. It is not a pretty time of year. Muddy brown colors cake the landscape covered by a thin layer of thawed soil that clings to boots, tripling the size of footprints. The vast fields that recently slept under several feet of snow have started to stir. But like a sleeper waking at dawn, the fields are foul-breathed; the refuse from last fall's harvest, last winter's kill, and the tons of manure mechanically flung onto the land have yet to be turned back into the soil. Their odoriforous juices mingle with the field's dying drifts of dirty snow where they will seep into the

226

softening ground for use by roots come July and August. The roads, no longer glued together by ice, show the wear of winter's forces too, cracks and holes and crumbling edges. The lanes carry the weight of large clods of dirt, flung from the whirling wheels of tractors now resuming their smoky commute between distant fields.

Down at the old Jenkins place, the tall grass that hadn't hidden Rudy well enough was flattened, sodden, and thickly matted. Months before, a snowplow had toppled the rusting mailbox perched in an old silver milk can. The For Sale sign was still up because the place was still for sale. The barn and outbuildings were empty. The milking chamber, usually the shining pride of any dairy farm, was overflowing with rotting wood, soggy, fallen insulation, and manure. Barn swallows darted through gaping windows. Pressed into the brown dirt of one hut floor, looking as if archeologists had just uncovered an ancient lesson, was the white skeleton of a starved calf, picked clean. Inside the ten-room house, dirty lace curtains fluttered in the drafts. The radiators were frigid to the touch. Every step echoed throughout the house. Everything had been taken from it except the memories and a discarded old board game called "Driver's Ed—A Safe Driving Game."

At the other end of Lincoln County, where it took nineteenth-century voters four tries before enough approved creation of a county named for the sixteenth president, George Briffett still leaned the snow shovels outside the front door of the courthouse each night, knowing how changeable April weather can be. Pretty soon the janitor would be filling the stone birdbath on the courthouse lawn. But for now, the edges of the huge glistening snowdrifts were just beginning to flow slowly across the busy sidewalks.

County Clerk Lee Smith had officially summoned eighty potential jurors to the three-story Ivanhoe courthouse from across the county's 540 square miles. Jury trials are rare in Lincoln County; it had been several years since the last one, some kind of a theft. Finding eligible jurors for a murder trial was no easy task near

spring planting time. Back at the turn of the century,
when Lincoln County had its last murder trial, it was
easier to find jurors, there being four families or so on
every 640-acre section, or square mile. But in 1980,
Lincoln County's population was only 8,207. Twelve
months later it was 8,168 and in 1982, 8,081. By trial
time it was about 7,800. In the middle of the Jenkins
trial, a new state study revealed that the number of
Minnesota farmers had fallen to 94,385 from a 1935
peak of 234,000. The study also confirmed what Kenny
Toft and other county workers had known for years,
that the number of medium-sized, 180- to 500-acre
farms, the economic foundation of many rural communities, was falling sharply. The number of small farms
was increasing as financially pressed farmers and their
wives whittled land holdings and took time-consuming,
off-farm jobs to support their country life. At the same
time, the number of large farms was increasing too,
many with out-of-state owners more concerned with
tax write-offs than soybeans.

The telephone lines crackled that rainy morning in
Ivanhoe, where the thirty-one-page phone book contains twenty-five pages of instructions. And all the
courthouse parking spaces on North Rebecca Street
were filled before 9:00 A.M. The big news was not the
trial itself; farm financial difficulties and tensions between farmers and their local financiers were old news.
In fact, a number of bankers had decided that they
really didn't need to make quite so many inventory
checks of collateral at isolated farms after all. Other
bankers stuck pistols in their glove compartments just
in case. Don Stokke, the sheriff next door in Lyon
County, had sent a couple of men out to cool down
one farmer who'd gone around saying that his banker
was about to become the next Rudy Blythe. The sheriff might have charged him with threatening bodily
harm, but that would only make matters worse. Don
was worried. "You have people here who've worked
this land for generations and someone comes along
and tells them it's all not been good enough, you got
to get out of here in sixty days. Well, that's the end of
everything they and their parents and their parents'

parents worked for all these years. And it's not 'cause they wasn't working hard. They was, until all hours of the night. It's 'cause somebody somewhere changed the interest rate so they gotta pay more. But they can't earn no more on their corn because somebody somewhere put an embargo on grain sales for political reasons and we lost some markets. And the fact is anyway, the more corn they try to grow to pay their debts, the lower the price goes on their corn. Now with these guys about to lose everything they've got, it all builds up inside, you know what I'm saying?"

The big news in little Ivanhoe (Pop. 761) was that the murder trial and the financial problems were big news elsewhere. A TV satellite truck parked out on the street proved it. So did three or four camera crews and the squad of guys with notebooks who didn't look like lawyers. Anyone who went over to the courthouse those days, and a lot of people seemed to remember some business they had to do over at the courthouse, saw the crowds outside the courtroom, including Blythe's widow and maybe even the kid himself. They saw the TV crews with their Japanese cameras and bandolier battery packs sprawled all over the steps, waiting and griping and swapping stories. They heard the gossip from the courthouse workers and maybe even noticed the new pay phone go up on the basement hall wall. That first trial night, and several other times during the three weeks of court sessions, townsfolk saw and heard Dan Rather talk about their town to the whole world. He even pronounced Ivanhoe right.

There was the usual sense of celebrity, exhilaration, and excitement, of being famous for a minute, and of hearing from faraway relatives who nearly died, they said, when they saw little Ivanhoe on the map, in color, behind Dan Rather's shoulder. But there was also a real sense of tension. Twelve people were going to have to pass judgment on this boy, on one of their own, in a real spotlight. Afterward, everybody else in the state and maybe some in the country were going to judge how well Lincoln County had judged the case. Now, talking down at the cafe is one thing. Everybody knows the rules, or the lack of them, and it doesn't really matter anyhow. It's just coffee talk. But getting up in front of a judge and all the cameras and people and reporters and saying, "Yessir, he did it," or "No, sir, this boy is innocent," well, now, that's quite another thing. Which is one reason why Swen Anderson did not want the trial moved out of Steve Jenkins's home county. It might just be a little harder to convict one of their own.

Forty-eight years and one day before the Jenkins trial began, the defense attorney, Allan Swen Anderson, was born in southern Minnesota, the first of two sons of a Swedish immigrant. It was not unusual in the Scandinavian ghettos that dot Minnesota, but Swen did not learn to speak English until the seventh grade. Critics of this eccentric attorney often maintain that he still hasn't mastered the tongue. All of which could well be another Anderson ploy. For Swen Anderson is no fool, though it may sometimes serve his purpose for people to think so.

Anderson commands attention, with his large, broad nose, his left eye in an eternal squint, an old-fashioned flattop haircut, and a Stetson perpetually plunked down over his flaglike ears. He can swear up a tornado of words that accidentally on purpose spill into nearby ears. "I got a voice you can hear in five counties," he says, suddenly smiling and then erasing it as his mind moves on to other things. Swen often talks only in capital letters, as in, "THAT FABEL IS A GOD-DAMNED PUBLICITY HOUND." The words are hurled over his shoulder with an accompanying stare just as the target passes by.

"Yeah, he's a character," admits his wife, Liz, who often serves as assistant and secretary in Swen's law practice in Granite Falls, about fifty-five miles from Ivanhoe. During this trial Swen would leave his motorcycle at home. After a near fatal bout with rheumatic fever and high school graduation at age twenty, Anderson earned degrees from the University of Minnesota and St. Paul's William Mitchell College of Law (with honors) before becoming an assistant to his county's prosecutor. He inherited his boss's job seven months later. For twelve years Swen was elected prosecutor, then somebody ran against him. Swen lost, which was fortunate in the eyes of some. Gruff, grumpy, disorganized Swen, with all his keys on a rubber band, strewing papers about his office as he rummages through files, doesn't seem to fit the mold of a methodical prosecutor.

Swen's love of underdogs led him into legal defense work, especially in defense of folks who were a little bit different, semi-outcasts in a small town where the unwritten rules can be strict because there is a manageable number of people to enforce them upon. These local pariahs, by some quirk of circumstance or taste or life-style, swim against the small-town current and get snagged on the law. And maybe some folks feel if this unusual fellow isn't guilty of that particular crime exactly, then he ought to be. That gets Swen's dander up and his energy whirling. When he takes on a client, he takes on the town too, if necessary. Swen usually gets his client off, and if the dog of the unpopular lawyer of the unpopular client comes home one day with birdshot pellets in the rump and the veterinarian won't treat the animal, Swen will nurse the little guy back to full health with a stubborn Scandinavian silence. Swen might even encourage the restored creature to run around town a little later, barking in capital letters and flaunting its recovery before whoever shot him to show that neither dog nor lawyer will be intimidated.

Swen Anderson took on Steve Jenkins as a client because he enjoys colorful cases and because another client, Louis Taveirne, was living with Steve's mother. And perhaps because some people were saying Steve was an oddball son of a sick father, neither of whom had done the farmer's cause any good, even if some bankers did have it coming and even if some folks overstated the link between a discouraged Jenkins and a discouraged region. And certainly because Steve had become a real underdog. Other lawyers were not exactly swarming onto the Jenkins porch to help Steve. "He had everything against him," Swen recalls.

Pending the trial, Anderson even got Steve out on

bail, $150,000 put up in the form of Louis Taveirne's local supper club. Swen promised the court that Steve would never leave his sight, which he didn't. The plainclothes police tailing both of them saw to that. Steve lived in the Andersons' home as another son; they already had four, plus an adopted daughter. Steve did odd jobs around Swen's storefront office just off Granite Falls' main street, where the walls are covered with the lawyer's personal trophies: a stuffed shark, some favorite fishing poles, guns, an original U.S. Mail pouch, and a portrait of an Indian, another underdog.

Swen has a great deal of empathy for his clients. He warns them about the opening of any trial when the prosecution unleashes all its big guns in a barrage of overwhelming oral observations and evidence so convincing that even an accused saint could come to believe he had done whatever they say. "A trial is like a pancake," Swen tells his worried clients and prospective jury members during courtroom interviews. "You can't tell if it's done until you see both sides."

A blunt Anderson also demands a great deal of discipline from his clients. He toughens them by hurling insults during long periods of practice testimony. "I like to call a spade a dirty shovel," he says, squinting again. Clients are not to talk to the press. No show of emotion in court. Do what you're told in public. Wear what you're told—long sleeves for Steve. It all adds up; Steve performs well for stern sergeants.

Early on, it seemed that Swen was prosecuting the deceased Jim Jenkins more than he was defending Steve. Swen had a handwriting analyst look at Jim's alleged suicide note. She found no indication of fraud and said that it came from a disturbed person. Swen hired a psychiatrist who interviewed Steve, his mother, his sister, his grandparents, and Louis Taveirne and then produced a fifteen-page report for the defense.

It found Steve to be very pleasant, courteous, cooperative, cleanly shaven, well dressed, above average in intelligence, and "neat in every regard" with "an exceptionally good moral code." He found Steve emotionally "overcontrolled" at times, except when the four-hour interview turned to his father, and Steve cried. The psychiatrist traced Steve's military fantasies to his father's tales of National Guard experiences, and he speculated that Steve, with his intense loyalty, obedience, and perfectionism, would have made an excellent marine, except that he seemed bothered by pictures of people being killed.

The psychiatrist devoted several pages to a profile of Jim Jenkins and his family based on the same interviews. He found him to be "a depraved maniac," overly dependent emotionally on his parents, especially his mother, with strong feelings of masculine inferiority and inadequacy, and an inability to function well at work away from his authoritarian father, who was disgusted by Jim's nomadic life-style and continued financial setbacks and had come to describe his son as a "deadbeat." The psychiatrist referred to family stories about cruelty to animals, Jim's refusal to take orders at work, violent temper tantrums in attempts to control people, notably his son, and a crying, begging emotional dependency on his wife in which he became, in effect, Darlene's third child. Jim also had repeated delusions about his wife's sexual affairs with men. Jim felt the banker was out to ruin him financially, the report stated, and Jim sought to emotionally enslave, subjugate and coerce Steve to stay with him and obey him, ending with a threat to commit suicide if his son left him alone in Texas.

The report found Darlene Jenkins to be the emotionally strong spouse, cool and controlled, in contrast with Jim's emotional flamboyance and insensitivity to others. Darlene was the family peacekeeper, the doctor found, and as a strong-willed female, could dominate the marriage and protect herself and her daughter. Jim's perceived inadequate relationship with his own father prompted him to direct his furies toward men exclusively. He said that Steve, on the other hand, was considered likable by everyone, cooperative with no disciplinary problems, and carrying a firm conviction that he must always respect his parents and never hurt anyone, even his father, whom he had come to see as sick and emotionally needy. Finally, however, Steve's inability to commit murder in a final shoot-out with police enabled Steve to make the emotional break with his father. In a final desperate act to hurt those left behind, Jim, by then a paranoid psychotic, committed suicide to leave a lasting emotional trauma on his disobedient son. The psychiatrist also stated that Steve, his mother, his paternal grandparents, and his mother's boyfriend all believe very strongly that Steve not only did not kill anyone but could not kill anyone.

During pretrial hearings that winter Swen appeared to be building a defense of duress, that Steve didn't do the shootings and had to be with his father that day out of his abiding respect for elders, concern for his own father, and fear for his own safety from a mentally-ill dad, and that too many doubts dotted the state's circumstantial evidence to put a kid away for life.

Thomas Lincoln Fabel saw this one coming before it turned the corner. Generally considered the senior and most experienced criminal trial attorney in the state attorney general's office run by the son of former Vice-President Hubert H. Humphrey, the thirty-eight-year-old Fabel is a rangy replica of a bespectacled Clark Kent. He has the kind of neatness and precision— the unwrinkled suits with striped red tie, the fine-point pens with fresh notepad, the crisp files, the precisely placed paper clips—that can quietly intimidate those who aren't always so ready and so organized. He must have everything he needs in those papers, otherwise how could he seem so self-confident and methodical? That was what the six-foot-four-inch lawyer liked so much about basketball and football during his college days at Carleton and before in high school—the plans, the precision, the team camaraderie, the psychological struggle of appearances, getting knocked down, knocking people down, and winning if you paid special attention to the details and thought quickly enough on your feet.

Like his legal opponent in this case, Tom is the descendant of European immigrants. His great-great-grandfather Philip left Hesse amid Europe's mid-nineteenth-century turmoil for the city called Philadelphia in the fledgling United States. From there, it was trains west and a steamboat up the mighty Mississippi to St. Paul, where Philip Fabel became bootmaker to the territorial governor. He launched his own local business: Fabel's Orthopedic Shoes. *FABEL'S SINCE 1856* said the downtown store's sign—until January 1981. Tom's uncle had to close the place down then; foreign shoes, national chains, and local malls were too much for one old family of hardworking shoemakers.

Tom had long before opted for a legal career and the University of Chicago Law School, just as he had long before decided that Jean Hoisser, his ninth-grade sweetheart, would be his wife. He is a staunchly religious man, perhaps not surprising for someone who has already died twice. At the age of sixteen at summer camp in the course of inflicting some mischievous midnight terror on younger campers huddled in their tent, Fabel fell off a tall, dark cliff and landed on the side of his head. Two times during succeeding days of surgery his heart stopped. Twice it was restored. His face and jaw required extensive reconstruction. The only sign today is a misbehaving hearing aid that sometimes goes beep in a quiet courtroom. Fabel holds hands with his wife and three daughters during grace before each dinner. Ralph, the dumbest dog on the street, rests underfoot. Tom likes to right wrongs, but realistically so. He's not out to change the world. He has seen too many victims and too many perpetrators who were once victims to think that all the chinks in society's dike could ever be filled. So much of what's wrong out there can only be shrugged at. Fabel just wants to clean things up a bit, fairly, according to his legal opponents, which doesn't mean he isn't tough. He likes the intellectual challenge of working within a binding legal framework that still leaves room for creative interpretations and arguments.

Although Fabel is one of the highest law enforcement officers in Minnesota, his second-floor office behind the state capitol in St. Paul is modest, overlooking the building's back roof: a metal desk, a single light, a mug that says *I LOVE ST. PAUL*, two potted plants, some file cabinets, and a phone, which he often answers himself.

On one wall is a large state map so Fabel can see where he is going to try the cases he regularly takes on to keep his hand in trial work and because he feels strange asking subordinates to do that work without handling some himself too. There is also a color photograph of a cabin in the northern Minnesota woods where Tom was headed on September 29, 1983, when he heard a radio report of a double homicide near Ruthton.

The other wall decorations in Fabel's office are drawings, mostly bright flowers, each crayoned by a different daughter—Leah, Anne, or Jessica. As he prepared for his long stay in a motel room near Ivanhoe, Fabel and his wife awaited the birth of their fourth child.

From his office, Fabel supervises five legal divisions from antitrust to public safety with twenty-three lawyers and fifteen investigators. Yet his presence out in the state is not always welcomed by the often part-time prosecutors who normally represent the state. Tom's presence means that something pretty important is going on. It also means that he and not the local man will be in charge, so Fabel must practice some tact as well as law. He must be aware too of simmering local resentments toward hotshot city lawyers coming to town to send a local fellow to prison. Resentment doesn't always show on the faces of jurors, but it can in their verdict.

"I admit," Fabel confided shortly before the Jenkins trial began and tension was building on both sides, "that some of my words do contain more than two syllables and you've got to be careful when a big-city guy comes to a small town. There are no more bumpkins out there. They are smart and savvy. And, anyway, no one can ever outcountry Swen Anderson. But I'm working on my speaking style. And I've got a family, too, and some boots to wear."

Thanks to the work of O'Gorman and Berg, BCA technicians, and other investigators, Tom also had a stronger case against Steve Jenkins than he had originally thought after seeing the boy's videotaped questioning the previous fall. After the detailed crime reconstructions and filmed reenactments, Tom had resolved his personal doubts over Steve's guilt. But convincing a jury that one of their local kids had brutally blasted two bankers into oblivion with some premeditation was going to require some skillful presentations of circumstantial evidence. He could place the kid at the scene; Steve admitted that in his grand jury testimony. But for a first-degree murder charge to stick, Tom Fabel and Mike Cable would have to do more than just place Steve at the scene. They would have to prove, beyond a reasonable doubt, that he had pulled the trigger and that he had planned to do so.

Tom had considered a plea bargain to get something for sure. As usual, he discussed this option with the victims' relatives. They were inclined to think that the justice system ought not be short-circuited by dealing. The trial should proceed.

The jury would have to be unanimous. Steve Jenkins would have to be the bad guy, not some mentally unbalanced father. And the bad guy could not be a fancy-talking Tom Fabel who came in from the Twin Cities with all his big college degrees.

Tom had always had great confidence in juries, especially their essential good faith and common sense. He would appeal strongly to those attributes. It often worked. But he was aware of the jury's unpredictability despite the fact that during the voir dire he and Swen could get a sense of each juror's individual values and feelings, or the lack of them, and each lawyer could eliminate the most objectionable, from each man's viewpoint. A murder trial is a visceral, emotional chess game between two lawyers using the law and twelve jurors who are closely watching, quietly listening, and then secretly dealing with each other's fears, hopes, and memories in a closed committee room.

Tom knew that few people in that area were rich. He knew that few had not had some dealings with a bank. He knew how much, generally, people resent needing financial help. He knew that deep inside the minds of some of those men and women waiting for their names to be called out in the courtroom, murder might be wrong, all right, but killing a banker was not murder. It was understandable. Rude certainly. Extreme perhaps. Not right, of course. But understandable. It would take only one example of such thinking on the jury and Steve Jenkins would walk out that courtroom door a free man.

When Fabel stands up at the prosecution's table at the start of a trial, he likes to be as well rehearsed as one side can be without knowing the opposition's exact plans, and he always remembers the trial lawyer's cardinal rule: never ask anyone a question without knowing the answer. So, like a good German Protestant, Fabel concentrates on the basics; flourishes can come later, maybe even spontaneously. For weeks before a trial, in a crescendo of concentration, he pores over the documents, the police and BCA reports, the photographs, the legal precedents for disputed points, the transcripts of interrogations and interviews. He compiles lists of vital points to be made and countered. He visits the crime scene and makes vivid notes so his realistic descriptions will place the jury mentally on the scene, perhaps helping to build his credibility, and so he won't be surprised by some defense twist. He orally rehearses portions of his presentation. He even scouts out nearby racquetball courts and potential players to keep the mind and body in shape come evening; stale tensions from long courtroom days, he finds, cause mental mold.

Tom would be helped by Cable, the quiet thirty-three-year-old son of an Indiana geometry teacher who would become the prosecution's designated Official Worrier. Another lifelong midwesterner who had seen the changes coming, Cable graduated from college and then worked seven long years in the steel mills of Gary, to finance his part-time graduate studies. Every evening, when the dark smoky sky would glow with the reflected orange flames of the country's largest concentration of blast furnaces, Mike would walk into a city classroom and study the law. Like Rudy Blythe and Jim Jenkins and thousands of other migrating Americans, Cable preferred the life of smaller cities and towns. So he studied the region, borrowed $600 from his father, and moved to St. Paul, where a three-month search for work ended with a job offer from a small law firm in Ivanhoe. When the county prosecutor's job became vacant in 1977, Cable ran and won. It provided some modest extra income ($17,200 a year regardless of the hours worked) with broad exposure to the area's problems and people, potential private clients every one. And thanks to the electronic wonders of satellite television from Chicago, he could also watch his beloved Cubs play baseball.

Mike could fret aplenty and he would have ample opportunity during the trial. His role in the case was not to speak in court, but to sit at the shiny, old wooden table in front of the judge's bench and speak into Tom's right ear, the one with the hearing aid, to register an observation or reminder or utter an encouraging word.

The first step in the trial was to pick a jury of twelve, plus two alternates, who would decide the fate of the kid whose face so many of them had seen before. Clerk Lee Smith drew numbered wooden chips from a box next to the judge and read off the number of the next juror to be questioned—Ines Krause, Karen Wagner, Leon Thompson, Wayne Jacobson. Seven of the first thirty-eight jury candidates disqualified themselves; they had known the Jenkins family somehow. Someone had tested Jenkins's milk samples regularly or had hired Jim Jenkins during haying season or said he or she just didn't feel capable of rendering an impartial verdict on the accused, though not in those words. Sister Cynthia Day was excused, for example, by simply saying her conscience would be bothered by participating. Mrs. Girard was excused because she had been Steve's fifth-grade teacher. Betty Solverg was Steve's Sunday school teacher. Mildred Hansen had hearing difficulties. Marlin Lustfield's son's ex-wife was Steve's first cousin.

Tom was very efficient and friendly throughout this crucial questioning. He could not be one of them, nor did he want to seem professorial or superior. Somewhere in between would be perfect, an uncle, perhaps, who came from a different background, who understood and accepted theirs, and who was outlining the rules and demands of a special experience they were about to share. He said parts of the testimony, the evidence, and the photographs would be shocking to some. One woman volunteered that she had once helped extract her son's wisdom tooth. "This will be a lot more grisly than that," said the prosecutor. The courtroom was silent as the message began to sink in. This was no *Perry Mason* rerun on the Sioux Falls station.

Through their questions, the lawyers sought to feel out these strangers' personalities. There were no right or wrong answers, although an alert juror would have seen Mike Cable or Swen's wife taking notes on their responses. Fabel, wearing his favorite gray suit and frequently touching the corner of his eyeglasses as if to adjust them, would introduce himself to each batch of potential jurors. "I'm Tom Fabel. I'm here to help Mr. Cable for the next couple of weeks." He talked about how important this trial was. His questions often reflected his interest—and underlying concern—with the jurors' attitudes toward the defendant, who was legally an adult and was about to be tried that way. Cable had warned Tom first thing about that: he'd heard a lot of talk locally about why the state was picking on a teen-age kid. "You know," Tom would say, touching his glasses, "in Minnesota an eighteen-year-old is considered an adult by law. Do you have trouble accepting that?" Some said no. Others said that might be a tad on the youngish side, while still others played it safe, saying they had seen some eighteen-year-olds who were adults and some who weren't. Tom would nod and maybe smile a little, and Cable would scribble something by that juror's name.

Both Tom and Swen were also concerned about hunters. Almost all the males in the countryside hunt to some extent, though not the women, which may or may not mean anything. Hunters are more likely to see nothing strange in a little boy pretending to shoot things until he got his own rifle for his tenth birthday, so Tom and Mike concentrated on ferreting out loners, making sure the hunters on the jury also belonged to clubs or hunted in sociable groups, not by themselves. One or two loners, identifying too strongly with a defendant like Steve, could deadlock deliberations or, worse yet, swing an impressionable jury to an innocent verdict with a plea for individual rights. Well, maybe he is different, but, hey, since when is that a crime in the United States of America?

Tom was also obviously looking for a leader or two on the jury. "You know," he would say to the intent throng of men and women, "a jury is kind of like a committee." And as Tom knows from sad experience, a committee without a leader is a hung jury and a large waste of time. In this regard, Dave Koster, the pharmacist from Tyler, seemed like a take-charge kind of guy. But Tom couldn't appear too pleased with him. That would insure the guy would be thrown right out when Swen's turn came to challenge selections.

Swen, for his part, was eager to line up two rows of red-blooded, all-American, National Rifle Association hunters who probably went through a stage when they played soldier every day and who had sons who did the same. "Have you ever hunted and killed animals?" he asked. Evald Evers said he hunted with a bow and arrow.

"Do you hunt successfully?" asked Swen.

"It depends on what you mean by successful," replied Evers.

"If you got one in the last five years, that would be successful," said Swen.

"I have yet to be successful," said Evers to a round of courtroom laughter.

Swen began by apologizing to all the would-be jurors. "I'll have to ask some blunt questions," he said. "It's not 'cause I'm nosy. My job requires it." He would mispronounce just about every juror's name the first time around, and maybe the second and third time. He'd apologize and fiddle with his pen and pad of legal paper, and put his little glasses on to check his notes and then rip them off from one side in a manner that drives opticians right up the wall. Swen would look across at the jury, sticking his head and chin out, raising his eyebrows, and pursing his lips. "It's very hard work up there," he would say during the recesses, but it looked as if Swen was just passing the time down at the grain co-op, apologizing first for asking too many questions and then learning some things.

"What kind of father did you have?" he would ask. "Did you obey him?" "Do you think grown men should occasionally cry?" Some did. Some didn't.

He was interested in their views on adulthood and teen-agers before pointing at Steve and reaching the $64,000 question for each potential juror: "Do you see before you now a totally innocent person?" Swen would freeze in his tracks, pursing his lips and sticking his jaw out again, awaiting the response. "Yes," they said softly.

Swen was interested in this Koster fellow too—the small-town pharmacist spoke out every answer clearly—asking him, as he did the others, "Do you have the courage to doubt?"

Koster right away revealed in his answer what his major concern was. "You're in a pickle," he said. "You could lose business if he's innocent or if he's guilty."

Slowly a general portrait of the jury began to emerge: they were people who got most of their news from television but claimed not to pay much attention to it; who all belonged to a church and its many affiliated social groups but preferred not to be president; who thought that a lot of eighteen-year-olds are really not adults despite the law; who often hunted; who felt obedient to their strict but loving fathers, who hardly ever cried but would have been forgiven for doing so if they ever had.

After two days of questioning and selecting, of trading vetoes back and forth, the prosecution, defense, and the judge reached agreement. They seated fourteen jurors, eight women and six men, including two housewives who wouldn't learn until later that they were alternates. There were two retired farmers, an active one, two nurses, a lab technician, a telephone repairman, and a former creamery manager. Dave Koster was on the list, too.

Tom Fabel had been worried at one point that not enough adults existed in Lincoln County who were free of bias and news media tainting to impanel a jury.

Now both he and Swen were pleased. In private, Tom had been describing Steve's developing defense as "Barnum & Bailey," focusing on "This Is Your Life, Jimmy Jenkins." In pretrial briefs, he opposed the handwriting expert as unscientific and the psychiatrists' report as blatantly biased since he talked only to those who had a definite stake in portraying Steve as pure and clean and Jim as a crazed maniac. Tom also sought to exclude testimony on Jim Jenkins's anger and strange behavior as irrelevant, since Jim was not on trial. He figured that much of the detail in Swen's briefs about Jim chasing some beer-drinkers off his property with a shotgun years ago or his racial remarks about a Filipino emergency room doctor one evening were put in the defense's legal papers, all of which became a matter of public record, not so much for the judge's eyes as for the eyes and ears of others. Such incidents might allay a widespread sympathy for the farmer, focus more on his craziness, and maybe plant some general seeds of doubt about poor Steve's guilt in the public's mind and especially in the minds of twelve TV news watchers and newspaper readers.

Just to keep things in balance so the press and the well-coiffed TV types couldn't get too caught up in this poor-farmboy routine and to alert the defense to how hard the prosecution was going to play, Tom did a little cultivating of his own. Along with his legal response to Swen's briefs, he included as exhibits some selected BCA investigative reports from interviews in Texas. No trumpets and news conferences to nail this kid, just lay a bit of the other side out there for the record to help the public perspective. These reports quoted Jim Jenkins's bosses and coworkers on how hard he worked, how reliable he was in the demolition work, and, in contrast, how strange and silent and consumed by guns and talk of violence was his son, Steve, who preferred to remove old windows with sledgehammers instead of screwdrivers. And they described Steve's attempts to build a bomb to drop in one woman's gas tank.

Tom's legal strategy also pointed out that under Minnesota law, for Swen to claim a defense of duress, he would have to prove that Steve faced the threat of imminent death from his father if he failed to comply with the man's wishes to help in the killings. And Swen would have to show too that Steve feared his father might kill him then and there in the farmyard, not later that afternoon or at some future time.

But during the Texas interrogation and three weeks later before the grand jury, Steve's description of the events leading up to and away from the killings made no mention of coercion. In fact, because of his father's night blindness and the need to flee in darkness, Steve ended up doing much of the driving to Texas while his father slept, which could prove cooperation, not duress.

So the duress defense appeared to be under considerable duress itself when Judge Walter H. Mann of Minnesota's Fifth Judicial District stepped from his Ivanhoe chambers to preside over the State of Minnesota, Plaintiff, v. Steven Todd Jenkins, Defendant. Mann looks like a casting agent's first choice for a district judge—average height, an open face, stern though ready to smile, and wispy, pure white hair, brushed back neatly until the wind blows.

The judge was a sixty-eight-year-old combat veteran of the South Pacific theater of operations in World War II and a thirty-nine-year veteran of courtroom combat, five of them served at the private attorney's table, ten at the county prosecutor's table, and twenty-four a few feet higher, on the bench. He had been appointed to the state court in 1960. Judge Mann is one of eighty-four judges who move from local courtroom to local courtroom in ten districts throughout the state. In 1978 he became chief of the five judges in the sixteen-county Fifth District, tucked away in Minnesota's agricultural southwest corner. Over the years he built a respected reputation for gentle firmness and fairness both among prosecutors and defense attorneys. Judge Mann's reputation was hardly affected at all when in 1980 the state supreme court publicly censured him after he confirmed a *Minneapolis Star* story that he had patronized a prostitute several times in 1977 and 1978. The judge was never charged with any crime, and in an explanation filed with the court, he described the paying relationship with the woman, a single mother, as affectionate and not illegal.

Judge Mann had heard all kinds of courtroom charges over the years, but the Steve Jenkins case was his first murder trial in many years, although a sudden unexplained death had struck very close to home. As he prepared for the trial, reading the documents and charges about the deaths of two family men and the suicide of another, he was reminded yet again of the death a decade before of his young daughter so far away, a drug overdose or a suicide, no one knew which.

On April 10, the first day of the trial, Judge Mann clutched a stack of books and documents as he mounted the steps to his chair behind the bulwark of blond wood and looked out over the pews full of spectators and reporters and relatives of the victims. He issued, in response to the state's pretrial motions, a set of rulings that shaped the proceedings and the form of the evidence the Jenkins jury would receive. He ruled out any lie detector tests of Steve, and the handwriting expert's personality assessment of Jim, saying the state of the art was too unreliable to be admitted as evidence. He also came down on the side of the state regarding testimony on Jim Jenkins's personality and behavior. The judge would allow testimony on the father's character only insofar as it bore on the son's actions. Anything about Jim's temper, for instance, would have to have occurred in Steve's presence or have been observed by a third party to demonstrate the nature of Steve's relationship with his father.

Swen expressed no shock at this ruling that devastated his defense plans. He rose and ripped off his little glasses and said he needed some clarification from the judge. One by one he listed a series of specific incidents that he wondered might be excluded. Would this decision, Swen asked, rule out any mention of Jim Jenkins chasing people with a shotgun as long as twenty years ago?

Yes, it would.

And racial remarks about a certain doctor at the time of Steve's spleen operation?

Yes, it would.

And cruelty to animals? How Jim Jenkins once deliberately ran a dog through a cornpicker or that he cut off the tails of twenty cows when one knocked his glasses off into the manure once during milking, but Steve wasn't present at the time?

Yes, that would be excluded.

Or how about when Steve was present and his father ran over an escaped calf with a pickup truck as an example of how the father treated naughty animals?

That would be excluded as irrelevant.

Well, said Swen, who seemed to be addressing the four rows of nearby reporters more than the judge up there under the American flag, what about the father's expressed desires in Steve's presence to kill other people, people like Louis Taveirne and Lee Bush? Would that be admitted?

That would be admitted, the judge said, if the defendant was present at the time.

Well, the same limits existed for the prosecution, didn't they, should the state try to introduce evidence about Jim Jenkins's good character?

Yes, they did.

The trouble for Swen Anderson was, however, that in most cases the only person who could testify about Steve Jenkins's presence at such a time and its impact on Steve Jenkins was Steve Jenkins himself. Although he hadn't announced it, Swen, for his own secret and strategic reasons, had no intention whatsoever of allowing his client to take the witness stand. Nor did the law require it. At the proper time the judge would instruct the jury not to read anything into Steve's silence. Swen, who was starting to act very fatherly toward Steve, rarely cared what anybody thought about him anyway, so all the whisperings and hallway suspicions that would follow this decision bothered him not at all. Only the real jury mattered, and Swen was going to do the absolute best he could with what he had.

Sitting next to him at the courtroom table was a dark young man, his black hair neatly brushed over his low forehead and his busy eyebrows shading small blue eyes. His hands were clean; he wore a small stone ring on his little finger. He kept his head down, looking at the table or at the legal pad where he wrote a few notes now and then with his left hand, rarely looking at the judge, the witnesses, the jury, or anyone but the Andersons and, during recesses, his mother downstairs in the hall. The clean-shaven youth wore nice sweaters and shirts, pressed trousers, and leather dress boots, not too well polished. Some days when the high school dropout entered the courtroom, his thin shoulders and 135-pound frame were covered with a varsity letter jacket that a few years before would have fit one of Swen's boys. If anyone had asked, Swen would have explained that Steve didn't have any clothes of his own or any money to buy a new, nonmilitary wardrobe.

Swen too had forsaken some usual garb. His beige Stetson was missing, a nod, he said, to the continuing bite of April's winds. Swen said he'd had a stubborn earache in his right ear and when he'd taken medicine, it had thrown his head off. So he stopped the medicine and pulled a large, formless stocking cap down over his big ears, as anyone past the nonsensical era of puberty would do in winter in Minnesota, where the football team is called the Vikings. The earache would cause Swen to cough many times during the trial, usually when Tom Fabel was speaking.

"Mr. Fabel, you may proceed."

"Thank you, Your Honor. May it please the Court, Mr. Anderson, Ladies and Gentlemen of the Jury. This trial is now about to begin."

The jury was impaneled, the spectator section was jammed, and television crews hovered downstairs. Tom began the prosecution's presentation by stressing how important and serious these days were for the individual jurors. He said there would be an opening statement by him and another by Mr. Anderson, although the defense could delay theirs until after the prosecution's case had been presented. Fabel called the opening statement a summary of what was to come, but he warned the jury that the summary was not evidence. Of course, all attorneys try to summarize their case accurately, he said, but just because a lawyer said some words didn't make them evidence. "Please remember that," the prosecutor urged.

Fabel said he would be calling a number of people to testify. The defense could cross-examine these witnesses and present its own case. But the defense was under no obligation to prove anything, the prosecutor said. That was the state's responsibility, to prove guilt beyond a reasonable doubt. Then there would be summations by both sides, instructions on the law from Judge Mann, and finally the case would be turned over to them for their decision.

"You can think of the evidence in a criminal trial as a bunch of little pieces," Fabel said, touching his glasses now and then as he walked back and forth before them. "It is not like TV. It doesn't come all choreographed and descriptive or necessarily presented in a logical or coherent order. Rather I think it's more helpful to think of the evidence in the case like a bunch of little pieces of a jigsaw puzzle that you spread out on the kitchen table. When you first walk into the kitchen and you see the mess the kids made at the table, you say to yourself, 'How can that possibly make any sense?' But then you pick up the cover to the box and you see the picture. You see the picture that the manufacturer intended for all those little pieces to make when they are fit together in a proper way. And that is what the opening statement is. The opening statement is the cover of the box of that picture."

Fabel outlined the specific charges. "Under our law," he said, "crimes aren't given nice clean names. Sometimes the same crime can cover two different forms of conduct. Likewise, sometimes a form of conduct can be characterized in two or three different ways." So there were six counts in the two murders. The first two were first-degree murder, both for the death of Rudy Blythe. One was for premeditated murder, the other for intentional murder during a robbery with a deadly weapon.

Counts three and four were second-degree murder, intentionally causing the deaths of Rudy Blythe and Toby Thulin. Counts five and six were felony murders, that is, the defendant intentionally committed a felony against the victims—assault with a gun or robbery—and that death was reasonably foreseeable as a probable consequence, though perhaps not originally intended.

Tom introduced the two victims, briefly described them, their family circumstances, their jobs, and how on September 29, 1983, they happened to be at the bank's farm, thinking they were meeting a possible buyer. "Instead, both of them were executed in cold blood. The question in this trial, Ladies and Gentlemen, is how did this execution happen. Who did it?"

Fabel described how Rudy Blythe's bank had loaned Jim Jenkins the money for his farm and cattle, how Jim Jenkins made all his payments until 1980, when he announced that his wife was leaving him and he was quitting farming, and the bank learned that many of the cattle had already been sold off. Then Jenkins changed his mind, said his wife was coming back and he wanted to try the farm again, but the bank didn't think he was a good credit risk anymore. So they took back the few remaining cattle. Jenkins declared bankruptcy, and the bank got the farm.

Then on September 28, Rudy Blythe got a call from a Ron Anderson. Rudy had never heard of Ron Anderson. But Anderson wanted to see the Jenkins farm. Ah-hah, thought Rudy, finally, a customer.

Now, the next morning at about eight, Jim and Steve Jenkins got in their pickup truck near Hardwick and headed for their old farm. In that truck, Tom said, they had a .30-caliber rifle, a twelve-gauge shotgun, a sawed-off .410 shotgun, some disarmed hand grenades, knives, and "other instruments of violent death." At eight-thirty they arrived. Keep in mind, now, they didn't expect Rudy for an hour and a half. They got out some guns. They took off one license plate. "Their plans went awry. For some reason Rudy and Toby decided to go out to the farm earlier than ten A.M."

Tom described Susan Blythe's morning, too, how she came to be at the farm, and how, when Rudy sent her for the sheriff, she last saw her husband alive fifty feet away walking toward the woods next to the garage and saw Toby, sixty yards away, walking by some old outbuildings. "Moments after Susan Blythe drove out of that driveway," Tom said forcefully, pointing to a large map of the farmyard, "Steve Jenkins began shooting at Rudy Blythe and Toby Thulin. Something happened. Something happened to get Toby Thulin from there back to the car in a big hurry. Something happened to get Rudy Blythe from in this vicinity into the car in a very short period of time."

Tom talked about where the bullets flew and what they did and how desperately Rudy was fleeing for his life, and how frantically, moments later, the Jenkins men fled too, stopping only to purchase more boxes of ammunition and to fire on a pursuing deputy. "That is a narrative account of the picture that is going to be on the front of that jigsaw puzzle," Fabel said. "Over against that account as part of the State's proof you are going to hear the defendant's story. It is going to be a very important part of the State's evidence."

Between Steve's questioning in Texas and in front of the grand jury the previous fall, the prosecution had a very detailed account of Steve's version. Methodically, Bob Berg and Mike O'Gorman had taken that version and laid it over Susan's account and over anything else they could find out about that day, those weeks, those two Jenkins men, and their thoughts. The overlay showed some crucial disparities, which Tom was going to drive home time after time.

Using simple declarative sentences, the prosecutor summarized for the jurors the Jenkinses' circumstances. They had returned to Minnesota from Texas. They wanted to start over once again with another milking operation. Obviously they needed money. They had difficulty. They visited banks all over that corner of the state, and everywhere the story was much the same. After examining their application, the banks would discover the earlier bankruptcy. They would learn that Rudy Blythe's Buffalo Ridge State Bank had been left with some unpaid debts in 1980. The other institutions would decide not to issue a new loan. But the Jenkins men continued to try, with mounting frustration. They tried to lease cattle instead of buying them, planning to pay rent with their milk checks. But no one wanted to extend any credit. Jim Jenkins blamed Rudy Blythe for that.

On September 28 came word from one last cattle dealer, who had seemed a good possibility: no credit. According to Steve's story, his father then set up the ambush, using the name Ron Anderson. "According to the defendant," Fabel said, "James Jenkins said, 'We are going to go out there to rob Rudy and to scare the hell out of him.' End of conversation, according to the defendant. According to the defendant, there was no further discussion of a plan during the balance of that afternoon. According to the defendant, they visited a friend's house. No further discussion of the plan. According to the defendant, they went out to eat that night. No further discussion about the plan. According to the defendant, they went through a normal evening's activity. No further discussion of the plan. According to the defendant, they went home that evening and went to bed. No further discussion of the plan. According to the defendant, they woke up on the morning of September 29, 1983, and his father said, 'Let's go.' No further discussion of the plan."

Fabel summarized Steve's sketchy version of the farmyard events: how they had been surprised by Blythe's early arrival; how Steve had run and hid behind the garage, not knowing where his father ran; how he heard some faint conversations, someone walking near him alongside the garage, a second car arriving and then someone saying, in a louder voice, to go for the sheriff. One car left. Then, Tom said, Steve recalled hearing someone say, "He's got a gun," followed by several shots, a period of silence, and more shots. When he hesitantly emerged from behind the garage, Steve said he saw a body hanging out of the station wagon and his father returning from the front yard with Steve's M-1 carbine. Steve said his father handed him the pickup truck's keys and said, "Come on, come on, let's ge out of here," and Steve drove away.

Tom recalled in a distinctly disbelieving tone Steve's story of the flight and how his father wanted to stop

and kill some other people and rob some places but
Steve talked him out of it; their arrival in Paducah,
Texas; and the end.

"Well," said Fabel, "that is the defendant's story.
And obviously that is not consistent with the picture
that the state says that it will prove. Just as obviously
there are only four people who were ever out at that
farm at the time the shootings took place, and three of
them are now dead."

Tom reviewed some of the discrepancies and im-
plausibilities. He noted how Susan had placed Rudy so
far past the barn that both Rudy and Steve would have
had to see each other if Steve had been crouching
behind it. He said the state would show how unlikely
it was that the Jenkinses had no further conversations
to plan their robbery and how unlikely it was that
anyone but Steve would pick up his favorite weapon—
which he admitted he even took to bed at night—and
fire it effectively.

Tom also made public for the first time how the
BCA had identified that funny metal noise Susan had
heard coming from the woods that morning as Rudy
dispatched her for the sheriff. Two weeks after the
killings, BCA agents had taken Susan back to the yard
and had turned her away from the woods. Mike Cummings ran around back there, touching and moving
anything he could find to re-create the noise. Finally
Susan's eyes widened. "That's it!" she said. "That's
the noise!"

Hidden beneath all the matted dead grass were some
old sections of metal gutter. Someone sneaking through
the grass had run over them, creating that muffled,
hollow sound. Susan often thought later that if she had
only made Rudy leave right then, if she had only said,
"Let's *all* go for the sheriff," he and Toby would still
be alive and she wouldn't be alone. By then, of course,
she had seen the police drawings and knew what that
sound meant: the killer was positioning himself to
shoot. He was moving from the back of the garage
around the chicken coop to the spot where agents
would find three empty M-1 carbine cartridges.

Steve had said that he did not hear any noise, but it

had come from just a few feet behind his hiding place. "If Steven Jenkins had been in this spot," Fabel said, "he would have had to hear the noise that Susan Blythe heard—unless that noise was made by Steven Jenkins. . . . That creates a real problem for Steven's story."

Tom said he would submit evidence about the shooting to show that Steve Jenkins was a good marksman, in good physical shape, and that his father, the only other man out there, did not possess the conditioning, the shooting ability, nor, ladies and gentlemen, even the eyesight to commit these crimes.

"That," said Fabel, shuffling his papers into a neat stack on the lectern, "is the big picture that we are going to get in bits and pieces over the next several days. On the basis of that, the state asks you to consider the charges you see before you. It will be for you the jury to determine whether the state has proven guilt beyond a reasonable doubt. Thank you for your attention."

Seventy-five minutes after he began, Tom Fabel was done. The judge cautioned the jury, as he would several times a day every day throughout the proceedings, not to discuss the case with anyone, read any newspapers, or listen to any electronic accounts of the trial. Ideally, the jury would have been physically isolated from any possible outside influences, bussed from a motel to court and back again every day throughout the trial or their deliberations or both. But the nearest motel was twenty-five miles away. And these folks had their own businesses and farms to oversee and planting time was fast approaching. And the defense had not requested such precautions.

During the recesses the jury, filing out of the courtroom first, would retire to a barely furnished room just behind the main chamber where a male or female bailiff stood guard, armed only with a stern face. Behind the closed door, many male jurors played cards, a game called Crazy Eights. They decided to bet a nickel a point on their running score during the trial, but by the proceedings' end no one remembered the scores and their minds were on more pressing matters.

Others read books, dozed, or chatted. Helen Renken did a prodigious amount of petit point.

Two doors away, Merle Mattson, the court reporter, could give his fingers a rest. For twenty years he has recorded everything in Judge Mann's courtrooms. But until the Jenkins case, he had never worked on a murder trial. Merle sat in the front by the bench all day every day, his fingers typing in a steady rhythm, translating the pain and gore, the legal phrasing and sterile ballistics into phonetic hieroglyphics to be transcribed and read later far away, as the official encapsulation of three weeks of courtroom stories dried and preserved for history, if it cared.

The television crews were prohibited from advancing beyond the second floor by a handprinted sign suspended from a piece of string. For the trial's first few days they lounged indoors on the lower stairs, talking, dozing, or reading. Whenever they heard feet clomping down the metal stairs from above, they knew a recess had begun and they sprang into position for forty-five seconds of work, filming Swen Anderson and Steve Jenkins descending to their basement workroom with blank faces, Darlene Jenkins on her son's arm. At noon they tried to convince Fabel to repeat for the microphones what he had just said at notebooks upstairs. That twenty-seven-second summary of his three-hour presentation could be spliced together with film of the other lawyer walking down the stairs and an artist's colored drawing of the judge looking down from the bench and rows of jurors looking on. Then came a longer sequence of the reporter standing in front of the courthouse at mid-afternoon when the light was better.

After the recess Fabel and Mike Cable called their first witness, Sheldon Thies, an assistant Lincoln County engineer who had made the enlarged courtroom drawing of the crime scene showing the location and relationship of various key buildings, the vehicles, trees, and the bodies.

Swen, who was saving his opening statement until later, was growling when he began his cross-examination. He questioned Thies closely on some measurements—how long the car was, for instance, and how far the trees were from the garage. Then after several questions he turned to the courtroom and the jury and stood silently a moment, nodding his head. "Thank you," he said.

Next came Sheriff Abe Thompson, who described the two phone calls that morning, his race to the scene, what he found, and how he identified the men by their driver's licenses and then photographed the bodies. Swen was on his feet with a foundation question to determine the sheriff's direct knowledge of these photographs (Thompson took them) and to inquire how far the sheriff was from the bodies when he took the pictures.

"I would say ten feet," replied the sheriff.

"Ten feet?" said Swen, almost drawing the question mark in the air.

Fabel resumed his questioning, asking the sheriff to describe the criminal investigative activities at the scene that day and how he initiated the hunt for the Jenkins men. Then, suddenly, Tom produced two photographs of the defendant, showing them to the sheriff for identification and then asking him to point out the same man in the courtroom, which Abe did. Tom wanted to know when they were taken (six days after the murders), if they were official county photographs (yes), and what the lines were in the pictures on the wall behind the defendant.

"It's a height chart," said the sheriff.

Tom then simply offered the two photos as state's exhibits, which brought Swen to his feet again.

"Your Honor," he said, knowing full well now what his opponent was up to, though few others in the room did, "we would have no objection as to reading into the record the height. We would object to the photos as to relevancy and we do not think they have any probative value."

The judge asked to see the pictures and heard Swen and Tom argue over their relevancy in an animated bench conference.

"The record may show," said Judge Mann, "that the objection is overruled."

As the two photographs were passed slowly down
the twin rows of the jury, the jurors could see for
themselves the same shaved commando with tattoos
who frightened Joyce Hall that stormy afternoon in
Texas. The eyes of most of the jurors went from the
image of the menacing youth in the photos to the
young man in the varsity letter jacket staring at the
floor by the defense table. Fabel was making a point
about this local boy without saying a word.

Swen's questions for the sheriff focused on details.
Were any rings or watches missing from the bodies?
Not that he knew of. Swen said he'd seen the victims'
clothing, but Rudy's socks and shoes were missing.
Did the sheriff know what happened to them? No, he
didn't. And by the way, were Toby's hands sort of
rigid and paralyzed? The sheriff said he didn't move
them.

The judge then said it was time for lunch. And
everyone left. Except that Swen and Tom and Mike
and Merle and the judge met privately in the back,
where, for the record, the defense attorney objected
again to the photographs of Steve as prejudicial and
vowed more strenuous future objections. This was not
the last skirmish in chambers that Swen would launch
—or lose—seeking to build a record for a possible
appeal.

Within minutes the Eagle Cafe, a former bar and
clothing store four blocks away, was filled not only
with the regulars from a handful of offices and businesses along the one block of Norman Street that
constituted downtown Ivanhoe, but with the chattering
groups of strangers from the courthouse. Inside the
purple restaurant, their nonstop conversations nearly
drowned out the frequent ring of the cash register.
You know, all those strangers might not be so bad
after all.

The judge and Merle were at one booth. The jury,
overseen by a bailiff, was at three adjacent booths or
tables. Spectators and reporters, who were really just
paid spectators though they got reserved front-row
seats, vied for the remaining tables and revolving
counter seats. It didn't take long for the newcomers to

learn some basic rules of survival: order the fast-disappearing slices of fruit pie first, and if today's soup was unappealing, ask for yesterday's; there were always leftovers.

During the first couple of days things got a little chaotic until everyone learned his role and the rules, and the bulky TV camera crews with their cords and camera packs and lights stopped strolling the two narrow aisles to record everyone's every bite. "It's always busy at noon," said Rhoda Larson, a waitress who was doing all right on new tips, "but now it's packed. And it's super, just super." The cafe remained full even though jurors were soon having their hot noon meals sent over to the little yellow jury room on the third floor of the courthouse.

Louis Taveirne ate at the Eagle too. And Swen. The prosecuting team, Fabel and Cable, was back in its courthouse workroom munching bagged sandwiches and reviewing the upcoming testimony with the two BCA agents. By a few minutes after one, they were back in the courtroom. At about 1:20 the spectators would flood in, some of them from more than 150 miles away. Susan Blythe came every day, often carrying a prayer book that she found provided increasing comfort, and set herself up on the inside end of the second pew from the back with Sharon Fadness at her side. "I want to know it all," she would say. "I have to know it all. I have to know what is happening." It is a reaction familiar to those who counsel survivors of violent crimes. Darlene Jenkins was sometimes at the other end of the same bench, sometimes one or two rows ahead, with her daughter Michelle.

Fabel in gray and Cable in black would walk in with a tall stack of papers and books. The prosecution and defense, though barely three feet apart, pretended to ignore each other and looked up at the bench, awaiting the judge. A gigantic old chandelier hung overhead beneath a pale round skylight to illuminate the brown stone and beige plaster walls near a painted rendition of the scales of justice, precisely balanced. Sometimes the voices of playing children, outdoors at recess in the warming weather, seeped into the courtroom. In would come Merle, and a businesslike Judge Mann, who plunked down his documents and looked over the players before him.

"You may proceed, Mr. Fabel."

"The State calls Robert Berg."

Fabel established the witness's credentials as a state detective and submitted a series of slides from a projector propped higher on a borrowed book, *Helping Your Alcoholic Before He or She Hits Bottom.* Up on the screen appeared an aerial view of the white Jenkins house with the red roof Lyle Landgren and Dick Austin had laid and a few maples with their brightly colored leaves. And the barn and a station wagon parked near the garage with the front doors open. Fabel had Berg identify the place and date and show where the bodies had been found and where some shell casings had been found. Other pictures showed the high weed conditions, different farmyard perspectives, and a close-up of the car, where something seemed to be hanging out of the right front passenger door.

There was no sound in the courtroom save the soft whirring of the projector's fan, until the next slide came on. A soft gasp flashed through the room and some women put their hands to their mouths. On the ground was the face of Toby Thulin, father of three little girls, in unliving color. His eyes were askew and half-closed. His mouth was open slightly. His hair was neat, but the white skin of his face was pocked with little red wounds from flying pieces of glass, and below his chin was the neat little hole, a darker red, with a trail of fainter color running into his collar and up to his ear.

Fabel and Berg pressed on with other testimony and photos, and came inexorably to the colored close-ups of Rudy Blythe's bloated body, front and back, only partially clothed by his bloodstained garments. Berg inventoried Steve's military equipment and weapons, including the .410 shotgun with a hunting knife welded to the front and some throwing stars with specially sharpened points. Because he had never heard of such weapons until this trial, Tom had Berg explain these Oriental martial arts weapons. Berg identified a wide variety of other items that afternoon, from bullet fragments from Toby's body to Rudy's blue boxer shorts and the pair of eyeglasses found in Jim Jenkins's jacket pocket.

Fabel seemed finished then, but Mike Cable whispered something in his ear. And Tom had one more question. He recalled how hard Berg had said it was to wade through the thick weeds on the Jenkins farm. And he thought some jurors might have connected that difficulty with Berg's obvious limp.

With a little smile that anticipated the upcoming courtroom snickers, Berg explained there was no connection. He had recently broken his ankle in a volleyball game. And Tom smiled too then. But not just over the image of a state detective falling down in a volleyball accident. The prosecutor, who had once alluded in court to his own hearing aid, had just shown that another member of the visiting prosecution team was human too.

Swen's long cross-examination focused on details. He wanted more information on crime scene measurements, and he wanted to know how much money Rudy had in his wallet (two or three dollars), if the spent shells might have been moved by weed removal (unlikely), and how frequent was the tendency of Steve's gun to jam. The prosecution claimed that only someone intimately familiar with this malfunctioning could have fired the murder shots. He also wanted to know if Rudy Blythe had any change in his pocket.

"Yes, there was a small amount."

"It is obvious, then, that Rudy Blythe had not been robbed?"

"Objection. Argumentative."

"Objection sustained."

What was the weather like that day? Swen wanted to know. Was it a little tough to see? Berg didn't think it was too bad, just a little slippery driving. What were all those weeds at the crime scene? Berg said they were just weeds. Where were Rudy's socks? Had they or any clothes been vacuumed by technicians? Swen seemed to be trying to suggest that Rudy had not walked past the garage, past Steve's alleged hiding place, looking for the trespassers; if he had, some of those weed fibers would have been on his clothing. But Berg, a veteran of many witness stands and a former U.S. Army MP, was not about to be ambushed, even by a momentarily smiling Swen.

Susan took her customary place in the spectator section the next day, but before the proceedings began Mike O'Gorman appeared behind her. He had seen her often after That Day, seen her turn from an overweight, distraught new widow full of tears and fears to a slimmed-down woman under emotional control, almost too much control sometimes. Mike bent down and whispered, "You don't need to be here for this morning's testimony." Susan, knowing what he meant, waited outside the courtroom for a few hours until Dr. Brad Randall was done testifying.

"Bullet Number One passed through the abdominal cavity," said Dr. Randall, who conducted the autopsy, "and, of course, the abdominal cavity contains the liver, stomach, and the intestines, and spleen." Using drawings and colored slides taken on his stainless steel autopsy table twenty-three hours after the bankers died, the forensic pathologist described in detail the wounds, fatal and otherwise, suffered by Rudy and Toby. Tom had to establish the medical cause of death for the record.

"This is a line diagram representing the bullet paths found in Mr. Blythe," the doctor was saying, pointing at the screen in the darkened courtroom. "Again, this is not a—"

The doctor's graphic testimony about shattering bullets and severed arteries stopped abruptly. He and everyone else had heard a kind of wheezing sound from the jury box. The lights came on quickly, and the judge dispatched Dr. Randall to examine a juror, Donald Guida. The elderly former farmer, who had a history of heart problems, looked pale and shaken. Following a recess, Judge Mann excused him from the jury.

Fabel called Cindy Sizer. She was the court reporter for the grand jury proceedings. Tom and Cindy read into the record lengthy portions of Steve's testimony from the grand jury transcript. Those portions described the Jenkinses' futile search for credit and talks Steve had five days before the shootings with Richard Hartson at his glass store. Steve had inquired what kind of bullet would pierce bulletproof glass. Then there was Steve's talk with Marvin Minette the day before the shooting about how to obtain dynamite. Steve had said he did not talk with his father about how they planned to rob and scare Rudy Blythe, and he described their movements leading up to their arrival at the old farm, how they parked there, got their guns out, and how he put on his camouflage shirt and floppy camouflage hat, and how his father picked up the M-1 and went to remove the rear license plate on his pickup truck. But they both heard a car slowing on the wet pavement of County Road 7. Steve said he ran behind the garage then. His father ran off too, but Steve didn't know where. Steve said he crouched there, hearing indistinct conversations, pauses, then three minutes later another car arrived, and there were more conversations with a loud command to go for the sheriff. Steve said he stayed crouched there, facing west, doing nothing. Then one car left. Seconds later, the shooting began.

Steve said that when he emerged following a second series of shots, he saw one guy lying there, sticking out of the station wagon, and then Jim Jenkins appeared wearing his usual blue jeans, denim jacket, and tightly rolled purple and white stocking cap and told his son to start driving.

Tom Fabel and Cindy Sizer continued to read from Steve's grand jury transcript about the men's flight, their shooting at the pursuing deputy sheriff, and the taking of some license plates from an old car on an abandoned South Dakota farm. Then Steve had said he asked his father a question. "Well, I had asked him what happened back there, at the place," Steve testified, "and then after we got going a ways there, he said about since he got these two that he might just as well keep going, take as many with him as he could." They stopped for gas only once, Steve said, and for food not at all, eating only the few cans of groceries that had fallen through a wet paper bag into the truck in the rain the night before the shootings.

Steve described the long inventory of weapons, when and where he had acquired the guns, the stars, the handcuffs, the machete, and why he had sawn off the .410 and welded a knife on the front (it looked more military). He said how much he had practiced shooting (a lot), and at what—cans, bottles, a refrigerator door, a big, old tree branch dressed up like a man, and a piece of wood about a man's height with a plastic milk bottle for a head.

Fabel had led the youth through memories of his father's interest in guns: Jim Jenkins had none. His father hadn't gone shooting or hunting in years. Bad eyes, you know. When his father said he wanted to kill Louis Taveirne, too, Steve talked him out of it, he said. He also talked him out of robbing a lot of gas stations for money along the way to Texas, Steve said.

"Were you afraid of your father?" Steve was asked.

"When he got mad," he replied.

"Did you love your father a lot?"

"Yes."

Fabel opened the next week of the trial with Susan Blythe, who recounted the purchase of the bank and general business dealings since moving to Ruthton as well as the events of that September morning. Then it was Swen's turn. Interrogating the widow of a murder victim before a jury while defending the alleged criminal is a tricky business. There may be some points to nail down, but it must be done carefully lest the poor woman collapse in tears and the defense attorney and defendant come out looking like bullies.

Swen tried to protect himself a little; he asked Susan if they had ever talked before (no). As usual, Swen had a lot of detail questions, running over the dates of the Jenkins bankruptcy, his financial record with the bank, whether Steve Jenkins had ever had an account there. Swen was trying to separate Steve from his father's bad relationship with the bank, to leave only the father with a murder motive of revenge. But Susan was not going to give this, uh, person defending her husband's assassin one inch, not even a quarter of an inch. She had felt the first faint stirrings of sympathy for Steve in recent days. But for this sly, funny-talking lawyer who would thwart justice, she felt hatred and contempt. Her son Rolph, who was in court one day, felt the same. In fact, in the boy's regular nightmares about the murder, Swen became the menacing figure, not Steve Jenkins.

So when Swen asked Susan if Jim Jenkins had frequent overdrafts, she didn't know. And the bankruptcy, was that in 1980? Susan thought that was approximately right. Did she know Steven's age? No. Had Steve ever had an account at her bank? Don't know. Had he ever done any business there? Don't know. Well, would a bank normally have customers under age eighteen? Why not? she answered. Her eleven-year-old son had an account there.

Had she ever seen James Lee Jenkins write? Excuse me? Did she ever see James Lee Jenkins write? No. Did Steve ever have any loans at the bank? No, no. Do you know or don't you know whether Steve ever had any accounts with the bank? I said no, Mr. Anderson. He did not? I said no.

The tug of war went on. Swen questioned her on her husband's weight. Once she had said 260 and now it was 230. No, said Susan, she had said 250 before and 230 now; she had found a weight chart. And his height—once she had said he was six foot four and now, Swen said, scratching for any inconsistency and reminding her she had been under oath, it was six three and a half. He was a big man, Mr. Anderson. Yes, but why the difference? Sometimes, Susan said, she says she is five foot seven and a half and sometimes she says five foot eight. Well, which is it? I don't know.

But Susan knew exactly what Rudy had worn That Day. She was clothes conscious, unlike her husband. Like many wives of such men, she had tried to sharpen his wardrobe a bit.

When she had first been told she would have to testify as a crucial prosecution witness, Susan was very confident and wanted no directions. As the time neared, however, her confidence wavered. So her steeliness strengthened. She didn't know everything that was coming from Swen Anderson, but she knew what she was going to give him: no satisfaction nowhere nohow.

Swen pulled from a paper bag State's Exhibit No. 88. "Tell me if you recognize it," he said.

"These are my husband's pants, which he purchased from L. L. Bean."

"Do you know what kind of fabric it is?"

"I would consider that to be khaki twill."

"Do you know when he bought the pants, where he bought them?"

"Yes. I purchased them from L. L. Bean."

"What is L. L. Bean?"

"It's a sporting goods place in Maine."

"What? Maine?"

"Yes. Freeport, Maine."

"Other than the sight of blood and the bullet holes, is there anything different in these pants today than when you last saw them on your husband September twenty-ninth on the Jenkins farm?"

"Mr. Anderson, I wasn't concerned with my husband's pants, I was concerned with him."

"Did you notice any difference outside of the blood and the bullet holes?"

"I don't think so. I don't think so."

At the prosecution table, Fabel was looking just fine, on the outside. On the inside he was incredulous. How could this magnificent opportunity slip by? Any self-respecting widow he had seen would be in hysterics by now, melting like butter in the heat of such a blunt interrogation and likely solidifying the antipathy of the jury toward the insensitive interrogator, especially a jury with a majority of females. But here were these two stubborn people sniping at each other; Swen even asked Susan if she wanted to put her husband's bloody pants away. Tom, who had just about had it with Swen's irritating coughs during the prosecutor's presentations, considered spilling his water pitcher to get a recess and to get to Susan. "Loosen up," he would have said, "just be yourself." But he didn't. It wouldn't have mattered anyway. She was being herself.

Swen tried to chip away at the credibility of Susan's

memory. First the weight, then the height, then the color of the pickup truck she passed on the way to fetch the sheriff. So, a polite Swen thought aloud, maybe that noise Susan thought she heard back in the woods was a rabbit or dog or something. How loud was it, Mrs. Blythe? How loud is loud, Mr. Anderson? Maybe it was not loud enough for your husband to hear? Maybe, Swen continued, Rudy didn't walk as far back into the woods as she remembered, not far enough to pass the garage corner. Swen gave her a quiz on all different kinds of weeds.

The next round of questions came from Tom Fabel, who was obviously trying to clear the air. Was the noise Susan heard consistent with the volume and the noise she heard two weeks later with the agents?

"Yes."

"Thank you. I have no further questions."

Piece by piece, Fabel and Cable continued their attempt to assemble a convincing portrait of guilt. Agent O'Gorman described the reenactment with Susan and fellow agent Cummings, who is about Rudy Blythe's height. O'Gorman established that when Cummings walked toward the trees where Susan directed him, he appeared to shrink as he walked down a slight incline, as Susan had described. That point was well past the back of the garage, meaning Rudy would have seen anyone crouching down there, even someone wearing camouflage. Swen sought to point out how the police hadn't measured Rudy Blythe's pants legs and compared them to Michael Cummings's.

Fabel showed a videotape with agents and policemen walking where the tragedy's characters were said to have walked and reinforcing O'Gorman's statement about the incline, plus showing how if Steve was indeed hiding behind the garage, he would have had to have seen Rudy walking by. By the way, asked Tom, after the reenactment did O'Gorman notice the agents having to pull weeds off their clothes?

Objection.

Overruled.

No.

Swen, mispronouncing O'Gorman's name, had more

questions about weeds. He thought if Rudy really had walked back there, he would have picked up some weeds on his clothes. Swen had some illustrative photos of his own, showing a man on the Jenkins farm wearing khaki twill pants with some burrs and weed debris stuck to them.

Tom objected. Was O'Gorman present when these pictures were taken? No. Did he know what month they were taken in? No. What year? No. Did he know the man? No. Did he know where that man had been before? No.

"We would object, Your Honor, on the basis of no foundation."

"Objection sustained."

Paul Bartz, the part-time local policeman, described driving down County Road 7 near the Jenkins place and glimpsing a flash of yellow in the brown grass. He slowed, and saw part of a leg. Making a U-turn, he saw a white pickup truck speed down the driveway and turn south. No front license plate. The driver he saw was an older man, mid-forties maybe, heavyset. The passenger was a kid, early twenties perhaps, dark hair, green shirt. Rear plate was black and white, likely Texas. Six digits, letters and numbers. He went back and checked the ditch. It was a man; he appeared to have been shot. Another driver came by and was sent for help. Funny thing about that timing. Bartz was late leaving work because of the thunderstorm. Normally he would have been long past by 9:10 in the morning, and the second driver, Jeff Schroeder, would have been sixty seconds too late to see the truck there.

Swen tried to shake Bartz's identification of the pickup truck driver, reminding him that under oath before the grand jury Steve had admitted he was driving. "That would not change my mind, sir." Was anyone wearing anything on his head? The driver was wearing a stocking cap.

Okay, said Swen. Now, what was Susan Blythe's emotional condition when she arrived? Hysterical and combative, said Bartz. He was unable to understand her, but he gathered she was worried about her husband and she mentioned Jenkins, something about goddamned Jenkins. Then Swen showed Bartz a copy of his police report. He'd written it the same day, September 29, his wife's birthday, the policeman noted. The report indicated that Susan had said her husband saw somebody running in the woods.

"Like I say," said Bartz, "she was pretty incoherent and hysterical. But yes, I got the distinct impression that he had seen something."

Swen had nothing further to ask of this witness.

O'Gorman came back briefly to reconstruct Susan's drive to phone the sheriff. Because she had passed the oncoming Bartz at high speed and because he noted the time, the agents could determine how long everything took from the time Susan left the farmyard until the body was found—three minutes and twenty seconds. Swen thought they had the killer running a mite fast, according to his calculations, which he got confused in going over there in court. And perhaps so too did his listeners, which was fine with Swen.

Jeff Schroeder was up next. He'd seen the speeding white pickup truck too. When he came over the hill in his car just before the Jenkins place, he'd seen a truck pulled over and Paul Bartz said to him did he know cardiopulmonary resuscitation and Jeff said no. So Bartz sent him to phone for help, but there wasn't anyone at home at the Hartsons'. Schroeder doubled back to the Petersons', where the Missus was washing her hair in the kitchen but let him in to phone. By the time he got back to the ditch, Bartz was administering cardiopulmonary resuscitation but it seemed to do no good. Looking around, Jeff saw a trail in the wet grass running from the side of the house across the front yard and down to the body in the ditch, the only remaining sign of Rudy's last frantic steps.

No further questions.

Now, said Swen, could Jeff positively identify the defendant as one of the men in the truck? No, but Steve fit the characteristics of one of them, the passenger. Did the truck cross the center line at all? Jeff hadn't had too much time to watch. The white pickup was coming awful fast. But the driver appeared to

have the vehicle under control. Had Jeff noticed anything about the driver to indicate he had difficulty
seeing? Not enough time to make that evaluation, just
a second or two at the most. He was also asked about
Susan's appearance and he said she was very upset,
not quite in control of herself. "Shock, so to speak."

Tyler Police Chief Dan Fischer was up next. Soon
after the murders, he'd given up on the midwestern
countryside and found new work—in Arizona. When
he arrived at the Jenkins farm that September day, the
ambulance crew was working on Rudy, and Susan was
leaning over him talking and she said something about
her husband and Toby showing the farm to someone.
It was at that point that Dan realized there might be
others around, and they went up into the farmyard
and found Toby. No pulse in the neck. Dan sealed off
the yard and the path through the tall wet grass toward the old shed.

Ron McClure told Fabel about getting the radio
message at 10:10 A.M. to watch for the white pickup,
which he spotted later that morning. It had no front
license plate. He whipped a U-turn and came up
behind—it was a white Chevy, all right—and he got
the Texas license number, KW3618. Two male subjects. Passenger dark, very short hair. The driver wore
glasses and a stocking cap.

At one point the pickup made a wide left turn onto
a gravel road, and McClure lost sight of it for a moment behind the tall corn. Then he saw the truck stop
and a passenger got out with a rifle. "And I knew
exactly what was going to happen then.

"I heard two or three shots. I watched him come
across. As I came into the intersection, he come across
with a gun toward me, and then I observed him get
back in the pickup. And as soon as he did that, I did a
power turn around and headed back toward them to
follow."

Swen wanted to know if there was anything besides
the missing front plate that attracted Deputy McClure's
attention, anything like, say, swaying back and forth
or across the road. No, nothing. Had he measured the
distance between the shooter at the pickup and where

his patrol car was? No. Swen produced a transcript of McClure's interview with the BCA on September 29. The attorney wondered how come McClure could say approximately thirty yards then. "I was shook on September twenty-ninth." And why hadn't he measured it to begin with? "Probably an oversight on my part."

Tom Fabel had a rebuttal question for the deputy, thinking also of the pending charges against Steve for shooting at McClure's car. "If the defendant had been successful in shooting out your tires, as was indicated by his grand jury testimony, would that have prevented your vehicle from engaging in a high-speed chase?"

"I would object to that," said Swen Anderson, "I think it's speculative."

Judge Mann agreed. "Objection sustained."

"Well, let me ask it this way. Is your vehicle capable of a high-speed chase if a tire has been blown out?"

"No."

"Thank you. I have no further questions." Another day ended.

Lynnette Thulin was on the stand the next morning. She had brought her three daughters to court, as Susan had brought Rolph, to see the justice system at work. When one of them in the spectator section became disturbed at the questioning of their mother and the memories, it was Susan who offered motherly hugs.

Actually Lynnette, who was a couple of hundred miles away at the time of the killings and had been separated from Toby for nearly two months before his murder, had little to offer the jury. Fabel knew this. But he also likes to keep a trial from becoming too cold and technical in its detailed and legal discussions of ballistics and wounds and times and measurements, so he always tries to have at least one of the victim's family members testify to remind the jury that these deadly incidents leave real people behind.

To justify Lynnette's appearance, Tom's questioning focused on Toby's physical condition and mental attitude toward confrontations. He wanted to show that Toby would probably not run from a physical confrontation unless there were guns involved and he didn't have one. This would be a delicate process for Tom. He had to touch on Toby's military involvement and his veteran's status without going into it too deeply. He knew Toby had lied to several people about his membership in the Special Forces. What Tom wouldn't learn until much later was that Toby was an airplane weapons mechanic, not a swamp-crawling commando throat-slasher. He had nothing to do with psychological warfare or teaching little Vietnamese kids or aiming tracer streams of lead death down from circling Gatling gunships or being a POW twice or even once, brutally tortured and wounded by evil Vietcong. The closest that that son of a World War II veteran with exciting stories of dodging kamikaze planes ever got to combat, Toby's official military records would show, was a safe and secure air base in Thailand.

Lynnette described her marriage for the jury as being in the process of reconciliation following an August separation. She said what close friends they and Rudy were, and she recounted Toby's numerous physical activities with local teams, the National Guard, and his own daughters and, of course, his military combat training.

Tom asked her if she'd ever seen Toby in a confrontational situation. Swen objected. Tom objected to the objection. The jury was ushered out. Tom argued to the judge that such questioning was important because it would help explain how such a large, physically fearless man could be last seen walking away from the car, and then found dead in it. Tom said this was important because it spoke to the credibility of the testimony before the grand jury of the defendant, who said the shooting began almost immediately after Susan's departure. Lynnette's fifteen years of marriage to Toby would provide insight into his attitudes during argumentative situations, Tom argued. But Swen's objection was upheld. The jury was returned.

Tom started on another tack, this one more defensive than offensive, although it sounded innocent enough. The night before Lynnette was scheduled to testify the prosecution team met as usual after some

exercise—O'Gorman regularly whipped Fabel in racquetball. Cable, the worrier, referred to an early BCA
interview with Lynnette in which she recalled Rudy
speaking of an encounter with Jim Jenkins at his old
farmhouse one time when Jim chased him off with a
shotgun.

This incident came as a bombshell to Tom, who
hadn't seen that report or had forgotten it amid all his
other cases. Such testimony would be a priceless gift
for Swen, introducing the crazy-Jim factor that Tom
had just gotten excluded and putting a gun in Jim's
hands much more recently than even the defense could
establish. The agents remembered the interview, but
they had never been able to verify the incident. As
professional detectives, it didn't sound right to them.
It wouldn't be like the Rudy they knew not to spread
that story around, so convincing was it about Jim
Jenkins's lunacy. But the agents could find no one else
who said they ever heard about it.

Confirmed or not, the story presented a strategic
problem for Tom. Swen had copies of all police reports, so he could be ready to pounce. Tom could
simply not call Lynnette to the stand, but he rejected
that suggestion; he wanted living victims in evidence
and he had already told the jury the other widow
would be there. Changing course might raise suspicions.

There was, however, another alternative, a bold
stroke but Tom liked it. It fit the classic legal theory
that if there is damaging evidence to come out, bring it
out yourself, then smile. The prosecutor would go
ahead and produce Lynnette. Tom would steam along
right behind her with an appropriate smoke screen to
throw Swen off. If the defense attorney recalled the
damaging though unverified Jim Jenkins story, Tom
would just produce his "it-doesn't-really-hurt" smile
and wade on through.

After the jury returned, he asked Lynnette if Toby
had ever had any training in physical combat. Oh, yes,
she said. When was he in the National Guard? She
listed the years. Now, Mrs. Thulin, said Tom, while
you were living in Ruthton did you ever have any
contact with the defendant, Steven Jenkins? No, sir.

Did you ever have any contact with any other member of the Jenkins family including James Jenkins? No, sir.

"Did your husband ever inform you of any contact that he had had with the—with either the defendant or with James Jenkins?"

"Objection, hearsay."

Perfect. A suspicious Swen was trying to cut off that line of questioning, just as Tom desperately wanted. Swen didn't know what was coming, but he didn't like the sound of it, which was the whole plan.

"Objection overruled," said the judge.

Oh, no, thought Tom. He'd have to make the next question even more objectionable. So he spoke slowly, trailing the bait ever so carefully.

"Did you ever have any discussions with Rudolph Blythe concerning either the defendant or James Jenkins?"

"Objection, hearsay."

Phew, thought Tom.

"Objection overruled. You may answer."

What? Tom did not want this to happen! For once, he wanted the opposition's objection to be sustained. Quick now, how could he qualify this, stretch it out?

"Yes, sir, I did," answered Lynnette.

"When did those conversations occur?"

"May I give the background?"

"Yes."

"Your Honor," interrupted Swen, "I would ask the witness be responsive to the question."

"Yes. The background question would be improper, Counsel. You may ask specific questions."

Tom was getting his phony smile ready. Time was running out, and he was hoping and talking at the same time. "When did your first conversation with Rudolph Blythe occur concerning the Jenkinses?"

"Okay. It was the second [job] interview that Toby and I had with Rudy. The board wanted to meet Toby, and we were to go with Rudy to the Kronborg for supper to meet with the board. On the way we stopped at the old Jenkins place and Rudy took us into the house. He said that he frequently would stop to check to see if anyone had been there to vandalize it or anything. And while we were in the house, he showed me the destruction in the bathroom and in the plumbing downstairs in the kitchen. And he told me to be careful when I was standing in the bathroom—"

"Your Honor," interjected Swen, "I would object again on hearsay and irrelevant unless it is linked to Steve Jenkins."

"Objection sustained."

At last, thought Tom. "I have no further questions. Thank you, Mrs. Thulin."

Swen could still unearth something in his cross-examination, of course. "Mrs. Thulin," he began, "I am just going to ask you one question. Did Rudy Blythe seem to have ill feeling against Steven Jenkins's father, James L. Jenkins?"

"I would say—"

"Objection. Calls for speculation."

"Objection overruled. You may answer."

"I would say he was cautious."

"I have nothing further."

When Tom Fabel rose to summon the next witness, Al Weathers, the prosecutor was smiling, this time genuinely.

Alvis Henry Weathers was a fifty-nine-year-old gunsmith from Texas with a sense of humor and a simple way with words that is common down there and striking elsewhere. "What do you do as a gunsmith in Brownwood, Texas?" Tom once asked him. The answer came back: "We go broke."

In his questioning Tom established Weathers's expertise with guns dating back to 1942, when the M-1 carbine was the officers' rifle. In fact, Al helped write the navy's training curriculum on the M-1 during his thirty-year hitch. Tom had Al detail the mechanics of the semiautomatic M-1, how it fired every time someone pulled the trigger, how the ammunition cartridge fit, and how the ejection process worked. Tom introduced a business receipt that showed that Al had worked on a feeding problem in Steve's M-1, back on February 15, 1983. He mentioned notes Al had made then that said the weapon should have gone back to the factory to repair ill-fitting parts that pulled out the empty cartridge after firing. He described how that led to misfiring problems, when an empty ejecting shell would jam in one position, blocking entry of the new cartridge. He showed how in the military, men are trained to rapidly overcome such occurrences. Al also looked at the seven empty shells found by BCA agents at the murder scene.

"Was there any evidence of jamming on any of the rounds?" Tom asked.

"One," Al replied. It had a distinctive scratch.

Swen's long cross-examination opened on marksmanship, how good a shot an expert is at a hundred yards, leading up to what an easy shot it would be for an expert like Al to pick off at that distance a man standing, oh, say, six foot four. He talked about what a cheap reproduction the murder weapon was and how it hadn't been properly cared for.

When Tom's turn came again, he focused on the sighting mechanism and how difficult it was for people like Al, with vision problems, to use peep sights like the murder weapon's. Also, isn't it true that while a six-foot-four-inch man might be an easy target at a hundred yards, a moving target is more difficult to hit?

"Absolutely."

Now, assume you're shooting at a moving target about ninety feet away. You fire two shots almost simultaneously and both hit the target within an inch or so of one another. Is that a good or poor shot?

"It would be excellent," came the reply, which pleased Tom, but then Al went on, "if it was intentional."

"If it was unintentional, sir, does that mean it would simply be random chance?"

"Random chance."

"And I assume there is as much chance of accidentally shooting yourself in the foot as of putting a second shot at that same spot. Is that correct?"

"Objection. It is leading and suggestive."

"Objection sustained."

Then Tom questioned Al about running and shooting and how both activities seriously compound the problems of accurate shooting—heavy breathing, blurred vision, disorientation. "It takes weeks to develop the stamina," Al said. The worse shape you are in to begin with, the more inaccurate would be the shooting.

Tom also sought to bandage some wounds Swen appeared to have inflicted when he got Al to say he didn't understand how this particular weapon could have killed someone, other than accidentally, because it was essentially inoperative and in disrepair. Al told Tom, however, that before certain parts broke, the gun would have been operating fine, except for that stubborn ejection problem. Tom would later produce expert testimony that those parts broke during state ballistics tests. Oh, and would familiarity with this gun aid in its operation?

"Certainly."

Swen thought it would be impossible to put two shots so close together if you had to manually cock the malfunctioning semiautomatic. Right, Mr. Weathers? No, not impossible, said Al, which Swen did not want to hear. "I must tell you this," the witness added. "Normally when you have a manual operation, your sighting is a lot better than it is from your semiautomatic."

You testified, said Tom, slugging it out with Swen question for question, that a person could shoot an M-1 accurately without sighting. Well, said Al, what he'd meant to say was it's possible that out of a number of strikes, two could hit close to each other. "That's like dropping a jar of pennies," Mr. Weathers told the jury. "You are going to have a percentage of heads and a percentage of tails. I couldn't testify to how many heads or how many tails it would be, but I will testify to the possibility."

Tom had been making so much about all the training necessary to shoot well, and Swen knew pretty much what Charles Snow, top sergeant extraordinaire, was going to say on the stand about all of Steve's training. So Swen wanted to nail something down. But he hit his thumb instead.

Wasn't it true, Swen asked Al, that a man trained on the M-1 in the National Guard in the 1950s would be prepared to shoot the M-1 carbine too? Yes, they were the same guns functionally, just a shorter barrel on the carbine. And has it been your experience, Mr. Weathers, that qualified National Guard shooters are good or poor marksman?

"It's been my experience," Al responded, "that the people trained by the National Guard were poor marksmen."

Swen then asked Al why this was. He had to get out of this one with something.

The training was not intense enough or long enough, Al said.

"Would the training have been sufficient to have hit a man at a hundred feet that weighed over two hundred pounds and was six foot four in height?" Yes, because it's an easy shot. No further questions.

Tom produced Roger Papke next. A crime lab technician, Papke's role in the trial was to talk about murder scene measurements and ballistics. He set at 266 feet—nearly a 100-yard dash—the distance that the killer moved from shooting Toby to shooting Rudy. He traced the paths of the bullets and showed slides of test firings at the Blythe station wagon. The agents had wanted to know where that first bullet, the one that shattered the windshield, ended up. It was in Rudy's lower back.

Swen noticed that this fellow Papke wore glasses. Was he wearing them when he fired the test shots? Yes. Any difficulty making the shots? No. Any shots go off the mark? No. No further questions.

James Lansing, also from the crime lab, testified that the holes in the victims' clothing were made by the bullets; the fabric contained traces of lead. He linked the empty shells and bullet fragments to Steve's gun. He described the test firings on Steve's rifle and how it often malfunctioned, jamming with that poor ejection mechanism. Through a series of timely objections, Swen prevented the jury from hearing this expert's opinion on whether someone familiar with a particular gun could easily overcome its tendency to jam.

Tom called other witnesses as the days passed. Sheriff Thompson was back to produce a piece of the noisy rain gutter. Steve Wurster, an ambulance attendant, was called to say that he had not moved Rudy's body during the forty-five minutes they had performed CPR on him. Swen didn't care about the CPR. He wanted the twenty-eight-year-old to describe, over Tom's objection, Susan Blythe's condition in the ditch. "Very upset," he said. What else did he hear her say?

"She—before we even knew what had happened, she had made the remark that Jenkins had killed her husband."

BCA Agent Michael Cummings was called to produce a telephone message slip he had found on Rudy's desk by the phone. Barbara Jean Winsel, a bookkeeper-teller at the Buffalo Ridge State Bank, took the stand to say she had taken that phone message while Rudy was on another line. In her handwriting it said, "Ron Anderson is on 1." Beneath that in Rudy's handwriting it said, "10 A.M." That was underlined.

Swen began his cross-examination as usual. "Ms. Henshall," he said.

"Winsel," she corrected him.

Swen wanted to know if she knew the caller's voice, if it had been muffled. No on both counts. Was the call connected to Rudy Blythe? Yes.

Jerry Ihnen, Rudy's former bank manager, was called, basically to establish how the bank came to own the farm. He said it was collateral on Jenkins's loans for machinery and cows. Everything had been going along fine, Ihnen said, payments on time until August 1980, when he'd received an anonymous phone call from a man asking if he knew Jenkins was selling off his milk cows. Somebody out there had it in for Jenkins, or thought the bank was getting screwed, or both. Jerry didn't know anything about it. When he called the sales barn, they confirmed it; thirty-two of Jenkins's seventy cows were on the block, and not for the good dairy cow price of perhaps $800 each but for the $400 to $500 slaughter price.

The sales check was made out to both Jim Jenkins and the bank, Jerry said. When the farmer came in to cash it, Jerry asked him why he'd done it. Jenkins said his wife was leaving him, and he wanted to quit farming and pay off the bank. A few days later he called Rudy and said the bank might as well come get his last few cows. He was leaving. The bank never accounted for twenty-five missing cows. It took possession of all the property later from the bankruptcy judge.

Swen wanted to know about Jim Jenkins's walk (normal), ability to drive (don't know), eyesight (he looked at things and objects real closely). Did Jerry have any reason to believe Steve Jenkins had anything to do with the missing cows? No.

Nancy Smith was questioned next about her cafe in Pipestone, the Mayfair, and two of her regular customers, Jim and Steve Jenkins. Over numerous objections by Swen, she said they would spend at least an hour there every day, eating the noon meal and sometimes breakfast and supper, too. Tom asked her about the men's normal behavior in the cafe.

"Objection, relevancy."

Overruled.

"They were just very quiet. They never spoke to each other nor did they speak to the waitresses."

Now, the murders occurred on September 29, Tom said. What had Nancy noticed about the behavior of the Jenkins pair the previous evening?

Objection.

Overruled.

"They happened to be in the cafe at nine-thirty. And I usually go over to lock up at about that time. We lock up at ten, but I usually go a half hour early. And when I walked in, they were talking."

"Was there anything about the way they were talking that struck you as noticeable?"

"Objection, leading and suggestive."

"Objection overruled. You may answer."

"They, you know, other days when they would come in, they sat with their back up against the wall and would just relax. That night when I came in, they were talking to each other, you know, hovered over the table." They were the only ones in the restaurant at the time. They talked all the time, too.

Thank you, Miss Smith.

"Miss Smith," Swen began, "did—was there any disturbance in the cafe on the twenty-eighth?"

"No."

"Any bad manners?"

"No."

"If you hadn't heard the next day that there were two killings, would you have remembered that evening?"

"About them talking together?"

"Yes."

"Yes, because when I walked in I said to the girls—"

"I'm sorry, I can't hear you."

"I'm sorry. When I walked in the cafe that night, I said to the girls that's the first time I've ever seen those guys talk. . . ."

"Do normally people talk to each other in restaurants?"

"Yes, they do."

"Do they normally whisper so others can't hear them?"

"Not on the average, no."

"I have nothing further." Swen sounded tired.

Tom called the next witness and examined him with a rapid-fire sequence of questions. State your full name for the record, please. Richard Glenn Hartson. Lives on County Road 7 across from the old Jenkins farm. Age fifty. Married. Runs a glass business in Pipestone, D & M Glass. Known the Jenkinses for many years. Father and son had a good relationship. Saw Steve out hunting, but never the father.

Last time he saw them, Dick Hartson said, was the Monday before the murders. They came in, Steve all in camouflage gear. While Dick was out in the shop working on a tractor door, Steve asked him what kind of glass they used in banks and Dick said bulletproof and Steve asked what it would take to go through one and Dick recalled a supervisor who said a 30.06 wouldn't pierce it and Steve asked then what about an armor-piercing weapon. Dick paused and said, "Gol, Steve, I don't know. I have no idea."

Hartson described too the night Jim Jenkins flipped

his truck in the ditch and Darlene Jenkins explained, "Jim's got bad eyes."

Under Swen's questioning, it developed that Hartson had talked with Jim on that Monday as well. Dick had said he hoped all would go well with Jim's new milking operation, you know, not have any trouble with the bank and all. That's when Jim Jenkins said he knew his wife had fooled around with his banker.

Tom knew his notes cold, and he wanted another crack at Hartson's conversations with Jim. Aside from that day, Tom asked, had Jim Jenkins ever expressed suspicions that his wife had been having affairs with other men?

Objection.

Overruled.

Yes.

How many?

Three or four.

When Jenkins mentioned Rudy Blythe that Monday, did he sound bitter? No.

Now, about the bulletproof glass, had anyone ever asked before what kind of ammo would pierce bulletproof glass?

"No, sir."

Swen wanted to make a point too. Had Steve Jenkins here ever talked to Dick at all about bankers or anything like that?

No.

Marvin Minette, a well-driller, was called next. He remembered meeting Steve Jenkins the day before the murders at a mutual friend's house. Marvin was out back drilling, and Steve, dressed like a G.I. in jacket and camouflage pants, came over to talk. Steve asked Marvin if dynamite was hard to get.

"What led up to that conversation?" Tom asked, knowing full well the answer.

"Nothing," said Marvin.

Then Steve asked about fertilizer bombs.

Swen wanted to know if that was the sum total of their conversation. Yup. Did Mr. Jenkins at any time say he was going to make anything? Nope. Did he say he was going to use explosives to hurt anyone? No.

Steven Shriver then recalled his friendship with Steve Jenkins, and how they would go hunting together and shoot cans and try for animals in the state park. Steve was a good shot, his friend said, though sometimes that M-1 would jam. Steve had a quick little movement of just pulling the chamber back, dumping out the old shell, and ramming in a new one. But even with that jamming problem, he usually hit his target on the first shot, Shriver said.

Tom asked him about Steve Jenkins's ability to run and shoot. Yes, he could. How far? Maybe fifty yards? No, more like fifteen.

Tom then read Shriver a similar question from his grand jury testimony when he had mentioned fifty yards. "Were those answers to the grand jury true?"

"I've never seen him run fifty yards, but it's possible."

Tom also sought to question Shriver about his friend's physical abilities based on gym class at school. But Swen won out on an objection.

Swen wanted to make a point about Steve Jenkins's taste in clothes, which Tom saw as ominous. "In the summer and fall of 1983," Swen said, "was it the style of some teen-agers to wear camouflage clothing?"

"Objection, irrelevant."

"Objection overruled. You may answer."

"Some did. The boots and stuff were a lot cheaper than a lot of shoes or work shoes would be."

As the trial's second week neared its end, fatigue was starting to take its toll on the legal drama's main players. Swen was increasingly mixing up names. As for Tom, when Agent Berg was recalled, Tom asked if he had participated in some test firing at the windshield of the Blythe vehicle on September 29, 1983, the date of the killings. There had been no test firings on that date.

Through tone of voice and repetition of certain details, Berg signaled the mistake to the prosecutor. "Test shots fired at the windshield on September 29, 1983? No, sir."

"Excuse me," said Tom, correcting himself and moving on. He was more interested in Exhibit No. 77 anyway, a pair of eyeglasses that Berg identified as coming from Jim Jenkins's body and which were tested later by Dr. Theodore Fritsche, Jim Jenkins's eye doctor. What Berg didn't say was how the BCA agents had found the eye doctor of a dead man who hadn't left any records on all of his regional wanderings. The method: sheer luck, the kind of occasional good fortune that makes detectives believe in Providence. Bob Berg had no idea who Jim Jenkins's eye doctor was. But he did need a real good explanation of Jim's eye disease, retinitis something or other. One day in Marshall, Berg happened to mention this to Lyon County Sheriff Don Stokke, who suggested the agent talk to a Dr. Fritsche, an outspoken local supporter of law enforcement. The doctor did give Berg a helpful rundown on the eye disease and, on the way out, Berg instinctively asked one more question: Did you ever treat a James L. Jenkins? The doctor remembered no such name, but he had a secretary check. And there, waiting, was the Jenkins file.

On the witness stand, the doctor said he treated Jim for the first and last time on February 26, 1981. He explained his records and notes for that visit including notations such as "Night blindness, never drives at night" and while in the National Guard twenty years earlier "at night person marching in front of him had to have a white T-shirt pinned on his back." Jim Jenkins told the doctor his mother had some eye disease, possibly glaucoma, and his father was a diabetic. The doctor's notes indicated that, wearing his existing glasses, Jim's vision was 20/200; in other words, "he had a severe visual handicap."

Dr. Fritsche's testimony was peppered with revealing references to small-town life today: how the first question asked of a new patient is how he will pay his bill; how other patients are sought through referrals; and how the doctor must watch his pennies, squeeezing two tests onto one sheet of expensive testing paper. The testimony about eye tests was also very long and very detailed; Tom wanted it that way, more scientific, you see. The details of Jim Jenkins's eyesight problem were crucial to fingering Steve as the one who did the skilled shooting That Day.

The doctor said Jim had the early stages of cataract formation, limited depth perception, and such tunnel vision that he would have to keep shifting his head back and forth and up and down to cover the same field of vision others get at a glance. The best eyeglass correction possible for him would be 20/70 in the right eye and 20/50 in the left. The doctor noted too that Jim repeatedly tried to peek during the tests—a little cheating to better the results—so his actual eyesight may have been a little worse than results indicated.

The final diagnosis was that Jim suffered from retinitis pigmentosa, "an out-and-out hereditary disease that consists of a triad of constricted visual fields, poor night vision, and ultimate total blindness in a relentlessly downhill course."

Now, doctor, what would be the effect of two and a half years' time on the visual abilities of someone with this disease? "I can say without hesitation that it would be the same or worse."

Dr. Fritsche recommended that Jim Jenkins take five thousand units of vitamin A a day and that he get dark glasses because, studies indicate, both might slow down the disease. The records also showed that within three weeks Jim returned to the doctor's office. He didn't like the prescribed bifocals—too impatient with the necessary head motions. He had gone back to his old single-lens glasses, which gave him corrected vision of 20/200. The doctor said that those glasses were the same ones that Agent Berg got from Jim Jenkins's body, the same ones Jim had so carefully folded and put away before blasting most of his head off.

Tom had one other line of questioning. It seems that, in part as a result of his military training, Dr. Fritsche was something of a gun fancier. He owns several rifles including, son of a gun, an M-1 carbine. What, Tom asked, was the doctor's experience with the effects of eyeglasses on the ability to use peep sights on such weapons? The doctor thought they were a real handicap to shooting.

Could that handicap be overcome? Oh, yes, sure. How? By practice, practice, practice.

Swen wanted to talk to the doctor about driving

with such eye problems. Swen was a little worried about the inconsistency in who drove the pickup truck away from the farm: Steve said he did; Paul Bartz said an older man, Jim. Juries get troubled over such inconsistencies, perhaps wondering in their hearts. If the kid lied about that, maybe he'd lie about other things, too. So it would be helpful here if Swen could link Jim's eye problems to mishandling a vehicle, since at least two policemen had testified that the pickup truck did not swerve.

Doctor, could someone drive as fast as seventy miles an hour on a tar road and stay on the right side?

"Yes, I think that would be no problem."

"Would the fact it was a little bit cloudy or rainy or foggy make a bit of difference?"

"Not a bit."

Shit! "How about meeting cars then two hundred feet away?" No problem. Farther than that? Yeah, sure, six hundred, seven hundred feet, no problem.

"Now, that's working with the spectacles Mr. Berg gave you, at least the left eye had a twenty/seventy and right eye was a twenty/two hundred?"

"That's correct."

"Which is the same as my left eye, by the way," said Swen.

"Objection, Your Honor. Object to that and ask that be stricken from the record."

"It is stricken from the record, and the jury is instructed to disregard it. Try to avoid making personal comments."

Would some colors be easier for someone with these eyes to see on a human? Well, he would have a definite loss of color appreciation. Swen didn't pick up on that warning signal. He loved this boy Steve and he was getting a little desperate, though trying not to show it. Swen didn't know much about eyesight irregularities, but even someone nearly blind would have to pick up on anything as bright as Rudy's yellow rain slicker. Would this coat here, Doctor, be easy or hard for him to see? Swen looked around the courtroom to make sure everyone saw the bright yellow jacket as brightly as he did.

As I say, said the doctor, his color appreciation would be limited. He wouldn't see the same fluorescent yellow there we see. On the other hand, moving against a significant background, he probably could see him up to 450 feet or so. Swen was on to something.

"Would the yellow be a better color than some others?"

"Interesting that you should ask, because the MPS scanner that we did, the fluorescence on [the coat] is the identical yellow, and he missed all of the yellow dots, so his appreciation for that particular yellow is somewhat diminished."

There was no audible explosion at that moment. Swen didn't look as if he had just stepped on a land mine. He couldn't. He went right on to the next question about peripheral vision. Tom and Mike were furiously taking notes, furiously hoping that their activity would also underline that statement's significance.

Well, now, someone in the National Guard, even twenty years ago, would have to have pretty good eyes. What would Jim Jenkins's have been back in, say, '61 or so?

"The National Guard has a propensity to change its eyesight requirements depending on the need, as do most of the services, and I can tell you this from experience, when they really need men, they let the standards slip. Now, '61 the Vietnam War was just beginning. They might have lowered the standards. It's difficult for me to say."

Would Jenkins have trouble walking in the woods? Oh, yes.

Now this rain gutter here. It was lying in some weedy area. "Would that be the kind of thing Mr. Jenkins could trip over, James Jenkins, with his eyesight?"

"It sounds like something anyone could trip over."

Swen, not a little frustrated and annoyed, tried the shooting angle. People with this eye disease could compensate for it, right? Yes. Well, would his vision be such that as a proficient shooter, he could hit a moving rat at fifty feet?

A moving what?

Objection.

Overruled. Rephrase the question.

"Assuming that he was a proficient marksman, pretty good, could he hit a running rat, R-A-T, like a mouse that's big, a rat, could he hit a running rat at a rather short distance like fifty feet with a twenty-two rifle?"

"I would have to guess at that—I've had the same experience, and I'm a very proficient shooter, and I've missed many rats at the dump, but it's a tough shot," said Dr. Fritsche.

"Could it be done?"

"It could be done."

How about hitting a running man?

Objection.

Overruled.

"The answer is yes. Mr. Jenkins could hit the target with his eyesight at a hundred feet."

"Could hit a running man at a hundred feet?"

"That's what I said."

"No further questions."

Tom wanted to do a little patching work. About Mr. Anderson's driving questions, how is it possible to stay in one lane with this disease? No problem, with the signs and painted lines and experience driving. "Driving down a road is really no indication of driving skill. I did it when I was eleven."

About shooting with this disease, do you agree with your earlier statement that it would be very difficult, if not impossible, for Jim Jenkins to shoot the weapon with any accuracy unless he was a proficient shooter himself?

"I do."

Thank you.

Next was Dr. Michael B. McGee, a forensic pathologist in the Quincy mold, who has handled some fifteen hundred gunshot deaths in his career. He confirmed the autopsy report and spoke about each bullet in the killings. The one through Toby's throat, he said, also carried with it a small piece of chrome it picked up when passing through the window vent. The windshield bullet was deformed by its impact with the glass and, flattened, continued on through Rudy's clothing and into his back. The bullets test-fired through new windshields came out flattened almost exactly the same shape as the tear in the back of Rudy's rain jacket where he was first hit. From the angle of entry into his back, the doctor could tell the victim was crouched over. He was fleeing or hiding from something.

He described the four shots out front, two to Rudy's back—both fatally tearing up his insides, one severing the spinal cord—one in the arm, and one through the jacket. He could also tell that the victim was moving at the time of impact.

Perhaps as a legacy of his prosecutorial days, Swen let his personal curiosity take over his legal mind for a moment. Swen was very interested in where all four bullets had gone. Perhaps, too, he was trying to make more fog or set up more blind alleys of doubt for the jurors to feel their way through during deliberations, bumping into each other and finally giving up. But this was not Dr. McGee's first testimony as a police witness. First-timers might miss opportunities because they are nervous or seize up on the witness stand or because they don't know what's permissible in court, but not experienced courtroom veterans whose technical expertise can scoot fast past an interrogator who doesn't know as much. Effortlessly, Dr. McGee seized the opening in Swen's questions to make quite clear for the prosecution how accurate the shooting was. During one response, in a single sentence he managed to squeeze in three references to a bullet that came very close to Rudy's head. Not the mark of an amateur shooter.

Swen Borresen described his blacksmith business then, and the Jenkinses' frequent visits for welding work. One time just before the shooting, he said, Steve came in wearing three grenades hanging on a chain.

"Three hand grenades?" asked a stupefied Tom.

Yes, said Borresen matter-of-factly, as if it happened often. He wanted the bottom holes welded shut.

What for? He said a table lamp or something ornamental. Who picked up the welded grenades? Steve

Jenkins. How did he pay you? Cash. When was this in relation to the murders? About one week before.

Swen Anderson didn't care about the other Swen, so Ted Beard was called, all the way from his appliance repair shop in Brownwood, Texas. Did you consider yourself good friends with Jim Jenkins?

"Yes, sir, I sure did."

Did Steve and Jim have a good relationship? Yes, they never argued. Did he ever see Jim Jenkins abuse his son or boss him around?

"No, sir, I did not. He seemed to think the world and all of the boy."

Did you ever happen to see Jim Jenkins around a gun?

Yes, sir, said Ted Beard. It was right after Ted came back from elk hunting in Colorado. Ted was cleaning his gun and Jim was watching and said what a nice gun it looked like and Ted handed it to him. Jim looked through the sight toward a neighbor's house for a couple of minutes, moving his head around behind the sight.

"He handed me the gun back and said, 'I can't see anything through it.' And that is the only time I have ever seen the man touch a gun."

And did you ever talk about guns with Jim Jenkins again?

"Yes, sir, probably a little later the same afternoon I asked him if he would like to go deer hunting with me and my son. He said, 'Well, it wouldn't do me any good to go deer hunting with you. I would just be in your way because I can't see well enough to deer hunt.' "

Ted told how Jim couldn't see well enough at work to use surveying tools and how he walked kind of funny, as if he was feeling his way with his feet. Then at night he always, and Ted meant always, carried a flashlight and shined it right down in front of his feet.

A reassuring Swen wanted to talk with Ted about the rifle scope and whether Ted was taller than Jim (a little bit maybe).

"I am not trying to trick you or anything," the attorney said. He wanted to show how each user must move his head and adjust a rifle scope for his own eyes, like surveying tools.

Ted said Jim was a careful driver after he got the truck. Before then, people would drive him or Jim would just walk all over, in his funny, slow shuffling way.

Wasn't it true that Jim was employed down there in Texas as a night watchman? Well, now, Ted didn't know as how it was an actual night watchman. They had burglar troubles at one renovation site and Jim did stay overnight there a few nights to watch over the fuel pumps and all. But it was well lit. Was he armed? "Not that I know of."

Now, about the hunting, said Swen, shooting elk and deer, would they be moving at the time? Ted had seen it both ways.

"Would the deer be a fast-moving animal?"

"Faster than I like for them to be."

Ted's hunting pal Jim Perry was summoned next. James L. Perry, Jim Jenkins's foreman, had vivid stories of his employee's poor physical condition. Jenkins was hired for cement work, Jim Perry said. They were putting in curbs and gutters for a parking lot over at the junior high school and Perry needed good, strong men to muscle that gray, hardening mud about in wheelbarrows and then to pack it into place while it was drying. But whenever Jenkins tried to lift the 150-pound wheelbarrow loads, he'd spill them or become breathless and then his lower legs would swell up. Jenkins was a hard worker all right, Perry could testify to that. But he sure couldn't handle much physical exertion.

As soon as Fabel began that line of questioning, Swen objected. The jury was excused. The two attorneys argued. Tom said it was crucial to show the "circumstantial improbability" that someone in Jim Jenkins's shape could do a 100-yard dash in the farmyard and then squeeze off four rifle shots right on target. Swen said the whole thing was quite prejudicial. The two attorneys did some test questioning of Perry. And then Judge Mann said, No, that kind of testimony would be ruled out. Moving cement about didn't relate closely enough to the alleged activities. Bring the jury back. Score another for Swen.

Working with Jim Jenkins, Tom asked Perry, did you notice his seeing ability?

"Yes, sir. He was looking at some blueprints in the basement, and he said everything was blurred."

Perry said Steve and his father got along well, and then he got into the gun stories. Doing his night-watchman duties, Jim Jenkins had been carrying a baseball bat and one night Perry loaned him a shotgun loaded with rock salt, but Jenkins returned it the next day saying if he ever used it, he would get himself killed.

There was one other time. Charles Snow, the main-tenance director, had a horse on his farm. It had been sick for several days, wouldn't get up. The vet finally said, best thing to do is put her to sleep. Jenkins and Perry and some others went down to the farm. Jenkins was standing over by the horse and Perry walked up to him with a .22-caliber pistol and asked if he would want to shoot it and put it to sleep.

The attorneys and judge conferred privately then. They all anticipated trouble with this line of answers. It was revealing about Jim Jenkins, but not strictly legally relevant. The last thing Tom wanted was to hand Swen grounds for a mistrial or a vigorous appeal because he defied the judge's early ruling on details about Jim. The questioning resumed, carefully.

"Did James Jenkins make any statement about in-ability to use a pistol?"

"Objection. Relevancy and hearsay."

"Objection sustained on the first grounds."

"Were there any rifles present at that time?"

"No, sir."

"Thank you, Mr. Perry. I have nothing further."

The last prosecution witness was Charles Ray Snow, who was on his best behavior. Did you develop any weapons expertise during your army career and two wars?

"Somewhat, yes."

Snow recalled being impressed with Jim Jenkins's callused hands and hiring him, and what a good deal it was, because even after work, Jim lived there at the maintenance yard and Snow got a night watchman for free. Then when Steve came along it was even better because after a while Snow started paying him for part-time, but the kid was really working full-time just to be with his father. Snow said the two got along well together, sometimes seemed more like a brother-to-brother relationship than father-son.

"Did James Jenkins appear to dominate Steven Jenkins?"

"No way, no."

"Was the opposite true in any respect?"

"I think so."

Did Snow ever talk with Steve? Daily. Did those conversations ever involve military subjects or weapons?

"Almost all the time, yes."

"Who would initiate those conversations?"

"It started—probably Steve started to initiate the discussion, and then later on because of his interest—and, after all, I spent more than thirty years of my life in the military—I was more than happy to talk to him about military experiences I had had. And he was exceptionally interested in them."

Under Tom's questioning, Snow described how their relationship developed, how Steve bought the M-1 carbine with his first paycheck, and how, because the kid was so interested, Snow taught him the military firing positions. At length Snow described the different positions for the courtroom. Snow also recalled telling Steve he could get more stability, more accuracy, by installing on the carbine a web sling, which is then wrapped around the arm and is especially helpful when firing in the standing position. Snow was pleased to see Steve practicing these positions at different times.

"What kind of shooter was he?"

"Superb. Excellent. He hit the can every time. Danced it right across the pond." From the standing position, too. Snow said Steve was an excellent student.

Had Snow ever seen James Jenkins use a firearm? Never. How about Jim's eyesight? He said he saw poorly. Lots of times, well, we had the two of them down doing some demolition work at one school, and

Snow gave Jim the blueprints so's he'd know where the doors was to be cut, but sometimes his measurements was way off. "I attributed that to poor eyesight."

Did you ever see James Jenkins run?

"Never seen him run. He was a hard worker, but he worked at his own gauge, but never in a hurry."

"Did you ever see Steve Jenkins run?"

"Oh, yes."

"Did he have any difficulty running?"

"No. He was a good athlete."

"Thank you, Mr. Snow."

Swen just wanted to review a few things, to make sure Steve was respectful of his dad, which he was, which is important in this day and age. Or ought to be. And check on the distance from Steve to that can in the pond. How big was the can? Beer can size. What else did they shoot at? Tree limbs, tree branches. Snow told Steve he could hunt all the rabbits, skunks, and armadillos he wanted. Swen wanted to know what an armadillo was. Swen said the M-1 must be a very accurate rifle. Snow said it was at that distance.

For Tom again, Snow said he remembered that Steve hit pretty much everything he shot at. Did he have a sighting problem? Well, not after Snow explained the proper procedure. Steve always wanted to put the front sight in the middle of the target, and Snow kept explaining that it should go at the bottom. They went over proper breathing, sighting, exhaling slightly, taking up the trigger's slack, and then squeezing off the round—the standard firearm training for all members of the military. "He seemed to have an excellent knack for being a soldier and firing a gun," Snow said.

That was it for the prosecution.

The jury was ushered out.

"Your Honor," said Fabel, "the State of Minnesota rests."

Then, as agreed in the judge's chambers, Swen made a motion to have all six counts of murder dismissed because the state didn't prove each and every element beyond a reasonable doubt. As for the robbery part, Swen noted there was absolutely nothing to show that anyone was robbed. No watches or money missing. No robbery consummation, maybe just a trespass, which was a misdemeanor.

Tom disagreed, not surprisingly, noting several points of law and evidence including, right off, the fact that Steve said they went out there to rob Rudy and, according to Statute 609.185 Subsection 3, whether it was consummated or not is irrelevant. He requested the dismissal motion be denied.

Judge Mann agreed with Tom. He ordered the jury members returned. It was time for the defense to present its case and for Swen's opening statement. The defense attorney quickly glanced at his watch, worn on the inside edge of his wrist to facilitate such unobtrusive looks.

"Normally," Swen said, "I think the perfect defense is to call no one." But to buttress his claim that there was reasonable doubt on all six counts, Swen was going to call a few people who would testify about Jim Jenkins's physical abilities, coordination, that sort of thing. "There were two people at the scene," Swen said, seeking to redirect the jury's attention, "and one of the two was not the killer."

Swen said he'd be calling Eric DeRycke, a very nice man, a local attorney, to talk about how good Jim Jenkins could see. Another name had just come up, something of a surprise, claimed Swen, who hadn't mentioned it before. Gene Abraham, Darlene Jenkins's only brother and Steve's uncle, might be testifying. He's been a highway patrolman in good standing for some time and he might say that he had seen James Jenkins shoot the patrolman's revolver back in 1980 or so. There'd be a private investigator talking about rifle scopes. "That's about all he is going to say," Swen announced, "but it's quite important."

Finally, Swen said, he'd have Darlene Jenkins, the mother, to testify on what her husband could do physically back in 1980 around the time of that eye exam that Dr. Fritsche described so well.

So, Swen said, those are the main areas. "Of course," he added, "in every case a defendant has a right to take the stand or not take the stand. And the fact that he does not take the stand is no inference of guilt. In fact, when they had the trial of Jesus, Jesus never took the stand. And that didn't make him guilty of anything. And the same is true here.

"One other thing," Swen added. "The next most important item will, of course, be final arguments where the prosecutor goes first. And I will go last, which means nothing as far as the order of arguments. But that's the way it will be."

Minutes after the trial recessed for that last weekend, Bob Berg and Mike O'Gorman were in a car speeding the 150 miles northeast to Brainerd, Minnesota. No one outside the prosecution team knew about it. Swen's opening statement reference to Gene Abraham firing his service revolver with Jim Jenkins had stunned Fabel and Cable, and it bothered the agents that they might have overlooked it. Nothing had ever been said before about the state trooper testifying.

What the agents uncovered before they got home just before dawn Saturday disturbed Fabel greatly. Maybe he wasn't as bothered as he sounded in that early Tuesday morning meeting with Swen in Judge Mann's chambers. But Tom was disturbed.

It was "very, very distressing if true," Tom said. But BCA agents had interviewed Gene Abraham, and Steve's uncle said he didn't know he was going to be a witness, he hadn't talked with the defense in six months, he never said he saw Jim Jenkins use his gun—in fact, he never saw Jim Jenkins use *any* gun. If all this is true, Tom said, and he didn't know if it was, then Swen had no good-faith basis for saying that to the jury. Tom told the judge he didn't know how to undo the harm. He knew if the prosecution had done such a thing, the state supreme court would deal harshly with him and any conviction would be reversed. Now, even if the false evidence was not introduced, the jury might speculate on how the prosecution maneuvered to deny the defendant this point.

Well, now, said Judge Mann, the jury had been instructed that evidence is only that which comes from witnesses and exhibits, not words uttered in opening statements. In fact, Tom had reminded the jury of that in his opening statement. Judge Mann didn't feel, objectively, that any harm had been done. What did Mr. Anderson have to say? Was he going to call Mr. Abraham?

No, said Swen. Mr. Fabel had some good points. You see, Swen had been told about this on Friday by Darlene Jenkins who remembered her husband saying it, and she remembered the incident too, shooting Gene's .38. "I guess I relied on that," said Swen, "and she may still be right, having remembered it." But the information was not true, and he'd be prepared to tell the jury the information was wrong, but he hadn't acted in bad faith.

Tom reiterated his concern. The judge said weighing ethical elements was beyond the scope of this trial. Swen offered to make a correction statement as long as he didn't have to say where the wrong information came from because, Swen didn't say, that same source was going to testify. Tom wanted any statement in writing so there'd be no dispute over phrasing later. Swen agreed to say whatever Tom wrote out, which he did when the trial resumed.

"Mr. Anderson," Judge Mann said to Swen, the jury, a packed courtroom, and two very calm prosecutors, "it is the court's understanding that at this time you may have an amendment to make to your opening statement."

"Your Honor," Swen began, reading from a piece of paper that did not carry his own handwriting, "I would like to correct my opening statement. Ladies and Gentlemen, during my opening statement Friday I told you the defense would offer testimony from Mr. Gene Abraham, who I identified as a state trooper, and said Mr. Abraham would say James Jenkins shot a service revolver. That portion of my opening statement was based upon erroneous information, and therefore I want to tell you that Mr. Abraham will not be asked to testify because he does not possess the information that I had indicated, and that part of my opening should be deleted. Thank you."

"Gotcha!" mumbled one agent in the back.

"You may proceed with the defense," said the judge. Eric DeRycke was called first. A Belgian immigrant, an attorney. In fact, he once represented James L. Jenkins. What type of legal work did you do for him?

"I defended Mr. Jenkins on a bad check charge." Score two points for Swen: the guy knew Jim Jenkins well, and Jenkins wrote bad checks, it seems. An unsavory fellow, maybe capable of murder, too.

Eric last saw Jim in the winter of 1981, when he got out of his car to walk through snow on an icy patch up the stairs to Eric's office. They were poor steps, too, not built right. But the man had no problem.

"Your witness."

Tom's turn to snag, or drop, fly balls from the defense's bat.

"Mr. DeRycke, you have had as a client over the years a man by the name of Louie Taveirne, isn't that correct?"

"Yes, I have."

What legal work did you do for him? Well, his divorce and other small matters. Are you aware that Mr. Taveirne lives with the defendant's mother? Yes, he was aware of that.

Now, Mr. DeRycke, do you recall being interviewed on January 24, 1984, by BCA Agent Michael O'Gorman? Yes. Do you recall that Mr. O'Gorman asked you to estimate how long ago you saw James Jenkins? Not sure, but maybe. Do you recall telling Mr. O'Gorman your last contact was four years ago? Don't know. Might you have said that?

"It's possible."

"Fine. No further questions."

"Your Honor," said the defense attorney, "the State would call Jim Lenz."

"The defense," corrected Tom.

"Yes, the defense," said Swen. "Did I say something wrong? I'm sorry. The defense calls James Lenz."

Lenz was a farmer once, but he had to sell out—sold his cows to Jim Jenkins, in fact. Now he had a repair shop and sold auto parts and welding supplies, but his back had been bothering him lately so he didn't work so much. Last saw Jim the spring of '81. Saw Jim Jenkins run after a cow, no trouble. Saw him operate a crane, too, takes a lot of coordination, crane-operating does. And as far as James Lenz was concerned, Jim Jenkins was as good a crane operator as he'd ever seen. Seen him do welding, too. Jim Jenkins was a perfectionist. And mechanical work. Saw him strip down a crane once to replace some bushings. No sight problems. No walking problems.

Tom's turn. Mr. Lenz, you are a legal client of Swen Anderson's, aren't you? Yes. Now, when did you see Jim Jenkins do this crane work? Back when he was in the construction business. Okay, now let's calculate backward on that, said Tom. That's two or three years on the Tyler farm and another four years up at Ruthton and he's been off that place since 1980. So all this work you saw him do must have been around 1973 or '74, isn't that correct? Somewhere in there.

Now, about the welding. Tom backdated the welding memories an equally long way.

"Did you ever see Jim Jenkins use a gun?"

"No."

Swen had another turn. Well, what was it you saw Jim Jenkins do in the spring of 1981? Lenz didn't see him do anything. He just stopped by the shop, wanted Lenz to build him something. Lenz forgot what it was exactly. Did he drive his own truck to your shop and back? Think so, yes.

Next up was Darlene Jenkins, the star attraction. As the media exposure accumulated, spectators flocked in growing numbers to the trial. Transmitted through the air as tiny electronic dots, the characters in the legal drama had been transformed into celebrities, polished into a distorted replica of themselves, seemingly always involved in dramatic confrontations, seemingly always speaking in twenty-second snippets, seemingly never needing to visit a bathroom. They had become somebody famous to recognize coming down the stairs during a recess.

Steve and his mother seemed to have been awarded the highest celebrity status, launching the most looks and igniting the most whispers. Some of the teen-age girl spectators, even if they weren't lucky enough to get a number low enough to get courtroom seats, would nonetheless hang around the courthouse, watching for Steve, then watching him and giggling and maybe saying, "Hi, Steve." Steve, maybe with a little wad of chewing tobacco tucked in his mouth, would continue to talk with his mother and Swen and maybe nod and say "Hi" back. Then he'd try to smother a smile.

That Tuesday morning when Darlene was to go on, the would-be spectators started lining up even earlier than usual, well before 7:30 A.M. There must have been more than a hundred people. When the bailiff went to open the door for the 9:30 proceedings, the crush was so bad that she got knocked down. A man fainted. The door got splintered.

The spectator benches were packed. Susan was there with her prayer book and Sharon Fadness. Many sat with their arms folded, listening intently. The jurors looked uncomfortable wearing their importance, sitting up straighter in front, subject to the gaze of their peers. A thin, warmer wind whistled through a window crack.

At 10:15, wearing a proper black skirt and a very white blouse with collar ruffles that reached for her chin, Darlene Jenkins took the oath. Swen had her dance quickly through her marriage, her son, her divorce, photos of her ex-husband on a driver's license and in the National Guard, and identification of his eyeglasses, Exhibit No. 77. He wore them all the time except sometimes in the barn when milking and when he was welding. She recalled how upset Jimmy was with his new bifocals, soon giving them up for regular glasses.

Do you fire a gun? Yes. What kinds? Darlene didn't know all the names, but she'd shot her brother's .38 and a .22 semiautomatic Marlin and her father's guns and shotguns and her son's M-1.

"How good a shot are you?"

"Objection, irrelevant."

"Objection sustained."

Now, is this an accurate depiction of your old farm? Yes. Did you at one time have a windmill there? Yes. Please mark where that was. Thank you.

And did your husband ever shoot at that windmill?
Yes, he shot at the tail. Why?

"I suppose to hear the noise, like we all did."

"You could hear the metal ping."

"Yes, you could hear it. It would zing and ping."

How often did he do that? Free time, maybe a couple times a week with Steve's .22. Did he shoot anywhere else on the farm? Yes, out back where Jimmy dumped some silage. We'd all shoot out there.

"Could you point out where you used to shoot on the farm?"

"I don't think it's even on the board, Swen."

It was more of a garbage hole and the rats got into it. They were hard to hit. They see you, they take off real fast. But he got some probably at fifty feet.

She recalled how after they left the Ruthton farm and moved into a trailer in Marshall, the whole family would often go out to Dwire's old gravel pit south of town together and shoot with Steve's .22 at cans of all sorts. There wasn't much else to do. The targets were probably a hundred yards away. Jim wore his glasses then and he wasn't a bad shot, even without a scope. In fact, he shot a dog once. He shot the kids' dog.

"I was very interested to hear you had shot the M-1," Swen said.

"Yes."

"How did it function when you fired it?"

"It shot all right. It had, it had a lot of kick, though. I didn't like to shoot it."

Mrs. Jenkins, you said your husband could shoot as well as you. "Could you hit a coffee can at a hundred feet?"

"With what?"

"With a twenty-two."

"Oh, yes."

"I have nothing further."

Now, Mrs. Jenkins, Tom began, you've been here in the courtroom every day of the trial, haven't you? Yes. You've been meeting virtually every day with Mr. Anderson, haven't you? Pretty much. You've been traveling to and from the trial every day with him? No, I travel in a different car.

"You are now aware just how significant this whole shooting ability is to this trial, aren't you?"

"Yes, I would say I am."

"Now, do you recall, Mrs. Jenkins, being asked some questions about James Jenkins's shooting ability way back on October fourth, 1983, by Bureau of Criminal Apprehension Agent Dennis Sigafoos?"

"At that time, when everything happened—"

"Excuse me, do you recall being asked by him back on October fourth, 1983, about James's shooting ability?"

"I don't recall."

Apparently Tom wasn't surprised that Mrs. Jenkins didn't recall, for he had in his hand a transcript of that interview. He read a series of excerpts, questions about her husband's ownership of guns (none she knew of), about his interest in hunting (never went), in target practice (not at all), and her interest in guns (so little that she didn't know what guns her son owned). He asked her if she remembered her grand jury testimony in this same room and how she had been asked the same questions about her husband and guns.

"Do you recall being asked this question: 'Do you recall seeing him do any target practicing?' And your answer was 'No.' Do you recall that exchange in the grand jury?"

"No, I don't."

Tom asked Darlene if she recalled Agent Sigafoos asking her about Steve's M-1.

"I don't know. I have no idea what it is."

"It's your testimony right now that you and Steven went out and shot the M-1 together, is that correct?"

"We went out in the alfalfa field right in front of our house."

Tom wanted to make sure Darlene knew what Jim was in the National Guard. "A transportation unit, isn't that correct?"

"He was also in heavy equipment."

He wore glasses when they were married, right? Yes. Tom asked if it wasn't true that her husband's eyes got worse over the years.

"I suppose. Nobody's eyes get better."

"His eyes became noticeably worse, did they not?"

"Not to me, they didn't."

Tom ran through some of Agent Sigafoos's questions the previous fall about James Jenkins's eyes. Darlene had said his eyes were poor. He couldn't see anything in the dark. They were getting worse.

"Do you recall saying that to Agent Sigafoos?"

"I don't remember. Was—"

"I'm sorry. There's no question right now. Thank you, Mrs. Jenkins. I have no further questions."

Next, Swen focused on clearing up some things. Jim had owned a shotgun at one time, but he sold it. After that he shot Steve's guns, but he never went hunting except for the rats. And she remembered Jim shooting her brother's .38, but that was a long time ago.

Kenneth Nordine was up next. As Swen promised, he talked about rifle scopes. One point was that a shooter could aim a rifle with the barrel without fancy sighting and be pretty accurate, too, certainly enough to get a man at a hundred feet. Nordine was also familiar with scope sights, having qualified on a sniper rifle as a policeman. He explained about moving your head behind one to get the correct position. Otherwise the target would just be a blur. Nordine talked too about his police investigative experience and how important it was to keep and study everything, even a victim's socks and shoes.

In his cross-examination Tom wanted to make sure that Mr. Nordine was getting paid for his services to the defense's cause. Yes. And Tom made the point that scopes have very little room for adjustment, kind of one-size-fits-all.

Swen wanted a few more questions. Was Mr. Nordine paid when he was a policeman? Yes. Did he ever testify as a policeman? Yes, maybe fifty times. Does his testimony differ because he was hired privately instead of working for the government? No, sir.

The defense rested.

Fabel was somewhat surprised. Nothing about the duress defense at all. Tom had been setting up for that. Swen didn't even get his favorite newspaper quote in, the one about how gun collectors are good Americans too. "Every police officer in the state," Swen liked to say in defense of Steve's practice targets, "takes target practice at a human profile." One reason Swen didn't get to use those lies was that Tom never introduced Steve's tree-branch target that was dressed like a human. Sheriff Thompson had it ready in the next building, but Tom never called for it. Maybe the subject would come up in the closing arguments, which Tom had to finish meticulously assembling that day. First would come some brief testimony by Agent Sigafoos, who had been urgently summoned by telephone from another part of the state and was en route by car at that moment to thoroughly rebut Darlene Jenkins's testimony.

Swen's cross-examination of Sigafoos was longer than Tom's initial questioning. The implication of the defense attorney's questions was that there was more to the Sigafoos interview of Darlene than met the eye, or the ear. Swen's inquiries weren't too clear on this issue. He mentioned one of Sigafoos's questions to Darlene that referred to James Jenkins's story of a Darlene Jenkins—Rudy Blythe love affair, a suggestion that no one took seriously. Agent Sigafoos said he took that to mean she thought it was preposterous.

"Your Honor," said Tom Fabel, "the State rests."

8

THE END

"We have now arrived at the final phase of this trial," said a very tired Tom Fabel, who was by then running on adrenaline. It was like those final exams back at Carleton College—a set time for the test, a long night of study beforehand, reviewing, remembering, organizing chaotic human events and thoughts into an acceptable, rational, understandable order to convince his audience, even though the only place such human affairs ever happen in any rational, understandable pattern is in the next morning's newspaper. Tom's secretary, Nancy Haley, had not seen the boss so intense in a long while. She had driven down to Marshall with another attorney for the trial's end. Tom put her to work on his closing argument with a typewriter borrowed from the motel office. Her fingers flew over the black keys until nearly dawn transcribing Tom's writing and notes. He wanted a neat closing statement, as neat as his desk top back in the office, the ideas arranged I, II, III with a common theme throughout.

Though it didn't show in Tom's notes on the little lectern in Ivanhoe, there was much more on his mind nowadays than just this trial to determine someone else's guilt. Tom was carrying a load of guilt too. He had been on the road a good deal in recent years, perhaps handling more trials than he really needed to as the boss. He loved the work. He liked the legal intricacies and their creative use, and he liked the team camaraderie. But a year or so before the Jenkins trial, late one Friday afternoon after a dash from yet another trial far from home, Tom had arrived at his

303

house to find the girls at a neighbor's. He had walked in the door and heard Jean crying—in the midst of a miscarriage. Their fourth child—maybe, finally, a boy— was gone.

Life had gone on after that, of course, and on the scale of social tragedies that prosecutors routinely confront, it was low. Tom didn't say much about it. But late at night in those motel rooms, with the shag carpet that had known so many bare feet and the extra lamp anchored in the wrong place to read his files most comfortably, Tom felt guilty. He was always in touch by phone, but he wondered if, had he been in another job, perhaps in another profession entirely, maybe he would have been home to help more during the pregnancy, to relieve some of the special stresses of everyday life that can accumulate and become so heavy with one parent absent. Those thoughts kept creeping in around the edges of Tom's concentration on his closing Jenkins argument. They were even more compelling than usual because Jean was pregnant again. And again she was at home alone with the three girls while Tom worked a couple of hundred miles away. On the coming weekend, the Fabels were moving to a new house, one with an extra bedroom, and here was Tom in his good-luck gray suit with the red tie, addressing another jury in another small town far away from home.

He thanked the jury, as he always does, for their diligence in this difficult experience. The next few hours were very, very important, he said, and he outlined like a friendly tour guide what would be happening—his closing argument, the defense's, the judge's instructions, and then their own deliberations.

"Every time I think of this final phase of the American jury system," the prosecutor said, "I am amazed once again at the infinite wisdom that is embraced by the whole process. For weeks now you have been sitting here doing nothing but listening. All the talking basically is being done by lawyers and experts, people that are familiar with the courtroom. All of us that are participants in this trial are people that have gone to law school, and we have taken tests and we have done everything else to learn about the law so that we can participate in the trial. And yet when that final, most important moment arrives at the trial, the case is taken out of our hands, and it's handed to you. The reason for that is that you people bring to this courtroom something that no school in the world can give any person, no training can give any person. You come to this courtroom with all of your lives of common experience, all of your lives as good citizens of this county, collective experiences which are as different from one another as the faces that you have. And together you are able to bring to bear upon this whole decision-making process one important quality: common sense.

"That is the most important thing that you will be asked to exercise today. Please don't forget that. Above all, the process of decision making in a court of law is a sensible, rational process. Your decision should be one which is guided by your good common sense."

Tom spoke in plain, commonsense sentences, not too long, not cluttered with confusing subordinate clauses, not, above all, condescending. Just simple statements of fact, as if he were describing the weather, for a while, and how the oncoming systems related. He ran down the six counts again. He had a suggestion on how they could work through this complex process, which he said was rather like instructions for assembling a home appliance. First, Tom said, look to whether the defendant's grand jury testimony was completely true. If they believe that, then the first four counts of intentional murder would be inappropriate. But he could still be guilty of the robbery and assault.

If, however, they found Steve's grand jury testimony not completely true, then they should decide next if Steve was the shooter. If they found Steve was the shooter, then they must decide on premeditation.

Tom ran through the evidence then, attempting to methodically destroy Steve's claims to credibility one by one, as a careful gardener removes each weed without disturbing the roots of his flowers. Tom's favorite tool that day was repetition. Jim Jenkins announces one day that they will rob and scare Rudy Blythe the next day. Nothing more is said throughout the day, throughout dinner, throughout the evening, throughout the drive to the farm. When the pair was surprised at the farm, Tom recounted from Steve's story, Jim Jenkins grabbed Steve's favorite gun and ran. Hiding behind the garage, the youth didn't know where his father was and heard nothing but talking, no gutters rattling, no arguments, no threats. Just the shots. And when Steve emerged he found the bodies.

Even if they believed "all of those things," Tom pointed out how Steve admitted having all the deadly weapons and helping in the felony, the assault and robbery. Mike O'Gorman was correct back in that Texas questioning; it is as wrong to help a felon as it is to do it yourself. When you go out to scare someone with all these weapons, Tom said, death is reasonably foreseeable.

"Now, I expect Mr. Anderson to argue to you," Tom said, anticipating his opposition, "that these are merely the toys of a child." Tom paused. He picked up the murder rifle from the pile of weapons in evidence. "Some toys," he said. Then the prosecutor looked over at the eighteen-year-old looking down at his boots, his left hand cradling his cheek. And the jury's eyes turned there too. "Some child," he said.

Appealing to the jury's midwestern, middle-of-the-road common sense, Tom emphasized the implausibility of much of Steve's story, giving a statement of Steve's and then repeating the words "Extremely doubtful." You've heard from many people—Ted Beard, Jim Perry, Charles Snow—how close Jim and Steve were. Would they come up with such a drastic plan as robbery on the farm and not talk about it? Extremely doubtful. If they were going out there to rob Rudy, does a person rationally rob someone by running fifty yards away with a highpowered rifle? No. Tom moved over to touch the shotguns.

"Now, if James Jenkins did indeed plan on robbing Rudy Blythe or just scaring him well, what would be the far, far, far more logical weapon for him? For him, a man who, as you know, has a very limited eyesight, for him who, as you know, doesn't use guns. Right there! That's what you rob and scare the hell out of somebody with. You do it from a short distance. You walk up to them and say, 'What are you thinking about doing, not giving us a good credit reference? I want you to get off our backs, and we want your money, too.' That's the kind of weapon you use for that. Not that [pointing at the shotguns]. Not that. And certainly not from the distance of forty to fifty yards away from where the car came in. That doesn't make any sense."

He asked them to consider the murder weapon's history. Who bought it? Who took it to Charles Snow and said, "Hey, look what I've got"? Who trained with it? Who carried it everywhere? Who was that gun's best friend? Steve Jenkins. Not James. That's common sense.

Is it likely that with a strange car arriving, Jim Jenkins, who couldn't see well, would grab Steve's gun and run off across the open farmyard he knew so well to hide by the chicken coop where the shooting began? Extremely doubtful. "One of the problems, of course, in a case like this, and it is a common problem in murder cases, is that there aren't any eyewitnesses left around to tell us whether that happened or not. All we can do is work on circumstances."

Now, what was it that brought Toby and Rudy back to the car so far so quickly? Steve said he heard nothing after Susan left, nothing except two series of shots. "Something had to put them in the car. Something scared those men. Someone said, 'Toby, get back here.' Someone said, 'Hey, what are those guns?' Someone had an argument. Something happened. Keep in mind the men that we are talking about. Rudy and Toby. Big, strong, athletic men. Are they going to run away from just trespassers? No. They saw something that scared them. There was a confrontation. Somebody shouted a warning. Something happened, Ladies and Gentlemen of the Jury. The evidence doesn't tell you what, but the circumstances tell you that something had to have happened to bring Toby back, to get Rudy into the car, and to cause them to try fleeing. Defendant says he heard nothing. Can that be true?"

Now, if the jury decides that Steve Jenkins is lying,

they must ask themselves a question, Tom offered. "Why? Do you lie because the truth is actually better for you than your story? Or do you lie because the truth is much, much worse? What does common sense tell you about that? If you conclude that he was, in fact, lying to the grand jury, you can use that conclusion itself to support any other conclusions that you draw as to who was the shooter. The very fact that he lied can itself be evidence that he, not his father, was the man who pulled the trigger and killed Toby Thulin and Rudy Blythe."

Tom detailed the fine shooting. Marksmanship, ladies and gentlemen, marksmanship. Three shots in the backyard, all would have hit their targets but for an intervening object. And physical exertion, the chase. Then four shots in the front yard, not random sprays all over the universe either. But four precisely placed shots. "And what did you hear about your two candidates for being shooters?" Tom recapped Steve's passion for guns and his marksmanship and everyone testifying to his father's lack of interest in firearms.

"Oh, but there was one exception," Tom said, "Darlene Jenkins. Yesterday Darlene Jenkins told you that shooting was a family hobby of the Jenkinses back in the seventies." He contrasted that with her earlier testimony to the grand jury under oath.

Tom detailed the physical abilities of the father and son, the running and then the eyesight. The young defendant fresh from his Texas training under a Vietnam veteran or the aging diabetic with a 63 percent field of vision in his left eye and only 20 percent in the right eye, his aiming eye? The youth who rarely missed his targets or "the man who had to feel his way along the street because he had difficulty walking?"

Then, as he talked, as he spun out how improbable it was that the overweight diabetic with no interest in guns could do the shooting, Tom got the flash of an idea. When he came to the part about young, athletic Steve in training, Tom, clutching the murder weapon, threw himself into the description and dramatically threw himself onto the floor and sighted across the courtroom. The only sound then was the happy yells of children down the street at recess.

"Use your common sense, Ladies and Gentlemen of the Jury. That's what you are here for."

Tom turned to the question of premeditation, the hardest point to prove since it is within the mind, and brains don't issue receipts for proof. Usually, when you think of premeditation, Tom said, you think of Agatha Christie, where someone plans a murder for months and months. Well, that's not the only kind. You can use those strange conversations of Steve's as one indication they were planning something, the conspiratorial talk at the cafe, taking all the guns out there, that could be premeditation. Think for a moment more carefully about what happened at the farm. Toby, the wrong man at the wrong place, is shot.

"The defendant ran over and checked to see who that human being was. He evidently figured out that it wasn't Rudy. Who knows. Maybe his father told him. Maybe he knew himself. Maybe he didn't care. But in any event, from the moment that he leaves here until the moment he arrives at that spot, there is one thing in his mind. He is pursuing a man as a hunter would pursue an animal. As he leaves the car in this direction carrying that M-1 rifle, he knows what the M-1 does. He has just seen evidence of a dead man. He has used it many times himself. He has got one thing in mind, and that is the death of Rudy Blythe. As he runs out to the sidewalk, he has that one thing in his mind. As he runs down the sidewalk past the house, he has that one thing in his mind. As he rounds the front of the house and into the front yard, he has that one thing in mind. As Rudy Blythe comes into his sight, Rudy running for his life, he has that one thing in his mind. As he pulls up in the shooting position and brings the rifle to his shoulder, takes a deep breath, and expels it, he has that one thing in his mind. And as he pulls the trigger on four consecutive occasions, he intended one premeditated act. That was the death of Rudy Blythe. It is not necessary that premeditation exists for any specific length of time."

Toby's death was intentional, Tom said, second degree. Rudy's death was carefully calculated, however briefly, therefore first degree.

Tom knew from the fog-inducing questions of Swen's cross-examinations that Steve's defense and Swen's final argument would rely on reasonable doubt. One hundred different lawyers would give you one hundred different definitions, Tom said. The judge will talk to you of that later. "But keep in mind always that what we are talking about is a commonsensical standard. Proof beyond a reasonable doubt is not proof beyond all doubt whatsoever. It is a high degree of certainty, and that is what the evidence in this case produces."

Tom had a story to tell then, a story that probably each one of the jurors could understand, even from this city fellow. The story, which Tom had been polishing for more than six weeks, dealt with his family— his wife and three little girls back in St. Paul. Now, Tom has to do a lot of traveling and, of course, he calls them regularly and, well, let's imagine an imaginary conversation. Tom asks his wife what the girls did that day. She says, Why, they went to school, of course. Then Tom says, Well, how do you know they went to school? She might think her husband had been working too hard, but she humors him. She says she got up in the morning with them as usual and packed their school lunches. Annie was fussing with her hair as usual. Jessica came roaring into the kitchen late and in a mess, as usual. Little Leah was still sleeping. Jean saw them run down to the corner and jump on the school bus. At three-thirty, Tom says his wife says, she saw the children get off the bus and Annie had a new library book and Jessica had a math paper and they went off to play before their mother found the uneaten carrots in the lunchpails. That's how Mrs. Fabel knew they were in school.

"Now, you see, my wife didn't see the kids get off the bus up in front of the school. She didn't see them walk into their classrooms. She didn't see them as they sat there for seven hours, but she knew they were in school. And if she had to base an important decision in her life upon whether they were there or not, she could do that with a good degree of certainty. With a high degree of moral certainty. She never saw them, but the circumstantial evidence led her to but one commonsensical conclusion. That's the same that happened in this case."

Tom wanted to touch on Swen's anticipated closing statements. There were hints of improper investigative work. You know, Rudy's shoes and socks, which were examined and then returned to the widow. Maybe the defense might claim that if Steve were such a super marksman, he could have drilled the pursuing deputy sheriff instead of shooting at his tires and missing. Well, Steve and Jim Jenkins were trying to disable their pursuer then, not kill a policeman and attract forty-five more patrol cars. "It is a red herring," Tom said, "and don't let it confuse you."

Just use your own rock-solid commonsense standards, Tom said. Don't be overwhelmed by sympathy or emotion. Remember that the word *verdict* like so many words in our language, comes from Latin, the words *veritas dictum,* a true statement. "That's what you are to issue—a true statement.

"I ask in the name of the people of the State of Minnesota that you perform justice, that you render true verdicts. I ask that you render verdicts of guilty on Count One, the premeditated murder of Rudolph Blythe, and guilty on Count Four, the intentional killing of Deems Thulin. Thank you very much for your attention."

Tom sat down then, exhausted and emotionally drained. The trial had been like running a marathon; just keep focusing on the next hundred yards, the next hundred yards, and three weeks later after countless reminders about the next hundred yards, it was finally the last hundred yards. The prosecutor sighed deeply. He took his glasses off and sat, alone, for two full minutes with his right hand on the bridge of his nose while the trial went on around him. Lynnette Thulin walked by Tom's chair and patted his arm and he patted her hand and neither one said a word. It would be several minutes yet before Tom would begin thinking of all the other things he could have said.

Judge Mann called a brief recess before it was Swen's

turn. The defense attorney was obviously impressed when he got up to speak. For two hours he had sat there in his boots and brown suit, rocking gently on the back legs of his chair, his face distorted slightly by the pressure of resting his head in his left hand. Swen seemed to have conquered his cough.

He had some notes to consult as he talked. They weren't in full sentences, just phrases, key phrases, and points. Then Swen would rip off his glasses and speak to the jury with force. Many times those words weren't in complete sentences either, but the emotion was there.

"You have just listened to one of the most competent prosecutors in the State of Minnesota do an excellent job—Mr. Fabel," he said. "And in my forty-eight years of life this has probably been a very important—the most important moment, and it will also be in this case your most important moment.

"Detail in this case is everything," said Swen. "In every case the state must prove guilt beyond a reasonable doubt. This means that you jurors can act as human beings and don't have to be like God, human lie detectors. It is not for you to need to see what do I think, can I read his mind, can I read that mind. It's for you to go strictly on the evidence. And if there is evidence of guilt beyond a reasonable doubt, convict. And if there isn't, acquit. In this case it will be our posture that the reasonable doubt is tremendous."

Swen talked about circumstantial evidence and how murky it could be and how all of its inferences must point toward guilt. "Let's start out," he suggested with a smile, "by mentioning the one thing which Mr. Fabel never mentioned at all. And I don't blame him, because it's the cross of his case. And that is motive. Motive." Swen said he saw all the motive for murder on the father's side—losing the farm to the bank, believing his wife had an affair with the banker, discovering the bank was putting out poor credit references, "which I suppose the bank has a right to do. I am not saying the bank didn't. But again that was hatred and it would create motive." Hatred and jealousy were terrible things, Swen said.

But the kid was only fourteen when the farm was lost. "And I don't know any fourteen-year-old boy that is going to cry if Daddy sells his milk cows. Kids aren't that gung-ho for milking cows. And [Steve] never had a checking account in that bank. He never had a savings account in that bank. Where was his motive? And he never saw the banker Blythe or the banker Thulin." This is very important, Swen said, because "people do what their heart tells them they should do." He suggested that Jim Jenkins's heart was full of hate. Swen said he had always been grateful that he did not hate. It's a terrible emotion, he said. "I think hate, especially the hate of the poor that is very, very submerged, and it's continuous. And it has been continuous for a long period of time," Swen said.

Swen touched on the ability to kill. He recalled Darlene Jenkins saying how her husband had shot his kids' dog. "If you can kill your family dog—most people can't—that shows something," said the attorney. And how about the boy shooting at the deputy's tires, which showed he was a good marksman and trying to disable the car, even if he did close his eyes. Swen dropped that last phrase in out of the blue. "I can tell you," Swen said firmly, ripping off his glasses, "that those two bankers, they may have been killed by someone with poor eyes, but they surely were not killed by someone with closed eyes. No way." Swen noted that if the boy was such a good shooter, he could have killed the pursuing deputy and that would have made more sense, since a live policeman could radio for help even from a disabled car. "But he couldn't do it," Swen said. "He didn't do it. And I think that is a solid inference the boy could not kill."

Swen also found it interesting that the kid didn't destroy any evidence, the guns and army helmet and stuff. He didn't dump them in a river somewhere. And he turned himself in. That shows respect and trust in the law and Swen said he liked that. He also liked the idea of the boy talking his father out of killing or robbing anyone else. "That should have some consideration," Swen said. "Unintent should have some consideration of the way he acted."

Now about premeditation. Swen saw none whatsoever in this case. If they'd gone to the farm for the sole purpose of killing Rudy Blythe, why take off the license plates? Why didn't they have any food if there was so much premeditation? And no money. And no gasoline. Not enough ammunition. "And all this ammunition," Swen added, "ugly as it looks on this exhibit, is in the unopened box, unused, and it didn't hurt anyone."

As for the vision, Swen suggested that was a prosecution smoke screen. "Mr. James Jenkins was not a blind person that needed a cane and seeing-eye dog." Sure he had defective vision and terrible night vision, but that doesn't matter, because all of the shootings were in daylight. And the eye doctor had said fog and clouds made no difference. Swen reviewed the icy sidewalk testimony and Jim Lenz's about Jim Jenkins running a crane real good. Concrete examples, not hot air, that's what Swen said he liked best in a trial.

Getting Rudy Blythe was not all that hard, the lawyer said. "It was an awful easy shot because there was no zigzag running." Swen suggested, considering the path of the bullets within the banker's body, that Rudy had stopped and turned to look at the shooter, which made it an even easier shot. He also quoted Steve's grand jury testimony that Steve and Jim Jenkins had shot together in Texas, and since the only gun they had in Texas was Steve's M-1, that was a good inference that Jim Jenkins knew about operating the M-1.

Swen knew he had a problem with Darlene's apparent switching. So he touched on that. He thought some prosecution witnesses had more problems on cross-examination. Anyway, maybe the contradiction was just a misunderstanding on what Tom and Darlene meant by targets. Darlene did testify, after all, that her husband didn't hunt but did shoot at the windmill. "And that's something that sounds like fun to me. I kind of would like to do that myself. But he hit it and he killed the family dog. And he did kill a rat."

Swen said he liked clear evidence more than just

testimony. Well, he didn't mean that witnesses intentionally lied. Both sides had been very honorable in that area. But witnesses can forget, or color or change what they remember. As for the Jenkinses' conspiratorial conversation in the Mayfair Cafe, "I ask you, if somebody was really planning to join into a robbery, would it be discussed in a public restaurant? And could they have been talking about a bad soup? Is that really testimony that proves anything?" So what if they were so quiet? You'll note the grand jury transcript is full of times when Mr. Fabel had to say to Steve, "Speak up, Steve. You're not talking loud enough."

Swen noted that Jim Jenkins had taken one, not two, green bath towels that fatal morning. The towel was to be a mask for the robber, Swen said. It was obviously a one-man operation. Of course the father would tell Steve to run and hide. Dressed like a Vietnam commando, Steve would stick out, too easy to identify.

On motive again, it probably wasn't really robbery. Bankers carry credit cards, Swen said, not cash. "Many bankers wear Timex watches." Jim Jenkins really just wanted to cause trouble for the hated banker, to screw up any sale of the house. If robbery was the motive, Swen suggested, Jim Jenkins would have robbed Rudy and sent the banker walking off across the plowed fields while the robber escaped. And then perhaps anyone else, whoever went out there to consider buying the place, he'd be saying to himself, "Is that farmer going to stick his gun in my face too?" Swen called that "a feasible theory."

Yes, sure, the kid did what he was told—drove to the farm, got the guns out, ran and hid. "Is that enough to be proof beyond a reasonable doubt of intentional aiding another in committing a felony where death would be reasonably foreseeable?" Swen's voice indicated he had an answer. "Every church in this area, every school, every civic group teaches people to honor their mother and father and obey them."

As for Rudy not seeing Steve allegedly hiding behind the garage, Swen, who was desperate to sow

seeds of doubt anywhere, suggested that the banker
hadn't gone as far back as Susan remembered. The
gutter noise would be far more likely to have been
caused by a stumbling middle-aged man with poor
eyesight, perhaps seeking his hidden son, than an athletic young man. Swen's idea was that Jim Jenkins
emerged from the woods with the rifle, that Toby saw
him and ran to meet Rudy at the car, and that Jim
waited to open fire until he had two still targets in the
car.

The shooting there was not as good as the state
claimed. Three shots splayed all over. The one through
the windshield was obviously aimed at Toby and missed,
hitting Rudy instead. The fatal one for Toby wasn't
superb marksmanship at all, Swen continued; it was a
lucky ricochet off the vent window chrome strip. Rudy
might not have been able to run very well with that
first back wound; the doctor said that. The pattern of
hits and near misses out front didn't scream supermarks-
man to Swen. It was erratic.

Jim obviously chose the M-1 over the shotgun because there was more ammunition in the M-1 clips. A
shotgun holds only five shells. A person with bad
eyesight would be concerned about having plenty of
bullets on hand for his anticipated misses.

The bunk about the bombs and bulletproof glass,
Swen said, was easily explained as a curious young kid
asking questions.

Swen obviously hated to bring this next matter up,
but he comes from the Mark Antony school of law.
"Susan Blythe I think is a fine person," Swen said,
"and I have no grudges against anyone." But he said
her testimony left some serious questions that contrib-
uted to all the doubt that Swen saw in this case. Susan
couldn't identify the truck she passed on the way to
summon the sheriff because Susan had said it was
green and the truck is really red. Susan remembered
only one person in the truck cab and there were really
three, Swen said. She had trouble judging distance at
one point. It was quite feasible that she misjudged
how far her husband walked alongside the garage. If
her husband really had gone behind the garage, his

clothing, the pants, for instance, and the socks, although they were missing, unfortunately, would have picked up those sticky burdock seeds. Swen would also suggest that X-rays of Rudy's cordovan wing-tip shoes might have shown something helpful for the defense. All part of this doubt business.

"When Susan got back," Swen continued, "she was hysterical and combative. I don't blame her for being hysterical and combative, but it could affect her ability to give accurate testimony and judge distances. And the only thing the State has to prove the boy was behind here, and somehow got chased out around here, is the testimony that Susan Blythe thought on the second visit some twelve days later that her husband had walked behind the garage. She showed bias. And she even showed a little vulgarity. And maybe that's understandable."

Wouldn't you think if Rudy really did walk behind that garage, and if Steve really was the cold-blooded killer the prosecution claims, wouldn't you think you'd have found Rudy's body back behind the garage instead of way out front? See, that's the problem with circumstantial evidence. It can be read so many different ways.

Now, there are some other problems with the prosecution's case, Swen suggested. Jim Perry taking his new maintenance employee with the bad eyesight and making him a night watchman. Swen skimmed over the discrepancy between Steve claiming to have driven the getaway truck and two other witnesses saying it was the older man. Swen said this made Steve more believable because the boy knew it could hurt him, yet he told the truth anyway. Swen thought the chase actually proved that Rudy was chased by Jim Jenkins, not the young kid. The boy could have caught the overweight, middle-aged banker. "I don't know anybody that's forty-two that's as heavy as Rudy Blythe and could outrun a kid. I have kids. My children could not only outrun me, they could do it many times over."

Swen said he'd tried to point out all the many areas of reasonable doubt using concrete examples and not just talk. The jury might think of more during deliberations when they got really critical, and that was just fine with him, Swen said. If they did see this reasonable doubt, then they must find Steve not guilty on all four charges concerning death.

The accomplice question was more difficult, he admitted. "One of the big problems with the lad being just barely eighteen and being under the dutyship of his father is that the more obedient the boy is in honoring his parents, the more difficulty he has in these situations. It is really very tough. It is really very tough indeed when, if you obey your parents, you are a criminal. And if you disobey them, you are a delinquent." It was going to take quite a bit of soul-searching to convict Steve on this one, Swen thought. Of course, if they didn't have any reasonable doubt, then they should convict him. But the jurors had committed themselves at the start to being honest and truthful.

Swen Anderson thanked the jurors for going through this terrible and difficult trial with him. He knew they were aware that this young boy's life was in their hands. "He is eighteen. And in your hands his destiny lies. And I can only hope that all of you will—I hope that none of you are offended at me if I have throughout this course done anything or said anything to offend any of you. Don't take it out on Mr. Jenkins because I am his attorney. Do what you can on this. Do it as honestly as you can. And I am very grateful that I had the privilege of talking to you. And I thank you very much."

Swen dropped his notes and glasses on the defense table as he walked back. Sweating, with baggy eyes, he turned and seemed to be addressing a friend in the spectator's front row. "I did my best," he said. Then, for a moment, Swen smiled.

That was at 1:10 P.M. At 1:15 Mike Cable made a private prediction to a friend: conviction on two counts of second-degree murder. Tom Fabel wasn't anywhere near as sure. He felt good about the overall case presentation, but he knew that he didn't know, never knew, and never would know how the strange chemistry of any jury works, let alone this one. His instincts told Tom to be bothered by this one. The feeling started the week before. There is always a visceral relationship going on between jury members and the witnesses speaking before them, as each juror through his own eyes and personal set of experiences seeks to judge the speaker. Susan was so cold and controlled, so downright haughty, Tom was worried that poor country Swen might benefit from some sympathy bouncing off Susan. Susan was a key witness. Tom knew what the judge was going to be saying about witnesses.

After lunch the jury had seventy-one minutes of instructions from Judge Mann, whose bad back was bothering him again. He warned them to give a true verdict and not to let personal opinions about the advisability of certain laws influence their thinking. He told them they were the sole judges of the facts and the sole judges of the truthfulness of witnesses and the weight given each. They were not to accept the testimony of witnesses blindly but to subject it to their own judgment and experience. "You may consider their attitude, their frankness or lack of frankness, their fairness or lack of fairness, their demeanor on the witness stand, their demeanor in the courtroom, their interest, if any, in the outcome of the action, their means of knowing that to which they testified, and any and all other factors which you deem essential." If a juror's recollection of testimony differed from anything the judge or either lawyer had said, he should abide by his own judgment. "Sympathy, passion, or prejudice must not in any way enter into your deliberations," the judge warned.

He told them about the two kinds of evidence: direct evidence, such as eyewitness testimony, and circumstantial evidence. In order to convict on circumstantial evidence, he said, the circumstances must all concur that the defendant committed the crime. "If the circumstances, no matter how strong, can be reconciled with the belief that the defendant is not guilty, then the defendant should not be convicted. It is not enough that the circumstances coincide with and render possible the guilt of the accused. They must exclude every other reasonable theory."

The judge reread all six counts for the jury and defined their terms. He reminded them that in the United States all accused are presumed innocent until proven otherwise beyond a reasonable doubt. "Absolute mathematical certainty" is not required, he said. But reasonable doubt cannot be manufactured from imagination, conjecture, or speculation. They should draw no inferences from the fact that the defendant did not take the stand.

He stressed that any verdict must be unanimous. They were to elect a foreman, examine the facts, discuss them, listen to each other calmly and dispassionately, and avoid any arbitrary position. They could change their views but should not surrender their honest conviction solely because of others' opinions or merely to return a verdict. "Remember also, the question is never, 'Will the State win or lose a case?' The State loses the case only when injustice is done, regardless of whether your verdict is guilty or not guilty." He then excused the alternate juror and sent the remaining seven women and five men into their little jury room.

At 3:41 P.M. the deliberations began in that yellow room on the third floor. The waiting began everywhere else.

"This is completely excruciating, absolute agony," said Swen as darkness fell on the back steps of the courthouse. The defense attorney was huddled there with his wife, Steve, and Darlene Jenkins. "I love this kid too damned much for my good. It's hard to be objective. I tried to put on a dignified proper defense. It could go either way." One week later Swen would be back in a courtroom. The next time would have nothing to do with murder charges. Swen would be in court then seeking, successfully, to formally adopt Steven Todd Jenkins, making him in the eyes of the law Steven Todd Anderson. Darlene Jenkins agreed to the step, she said, because she felt the boy never really did have a father before. Twenty-one months later, Steve would lose this father too; on February 2, 1986, Swen Anderson, at forty-nine, would die in his sleep.

In a short time, the wait for a verdict became a long

one. The Eagle Cafe sent over the jury's supper and they worked right on through the meal, being careful not to spill gravy on their notes. First off, Dave Koster was elected foreman. It seemed natural, and he didn't resist. Everyone in the room was uncertain how to proceed. Only when the bailiff closed the door on their deliberations did the full seriousness of their task descend on them.

Dave naïvely thought that all twelve jurors would have their minds made up to begin with, and one or two votes would end the whole ordeal. Over the protest of some, Dave did hold a vote right off. The results settled nothing. In fact, the vote was just the beginning.

A few feet away, a score of people began their verdict vigil, when rational thoughts give way to hidden fears. Relatives, reporters, and local police lounged around the courthouse, quietly swapping stories, excited at first and then quickly bored by the wait and the hard seats. The hours ticked by very slowly.

Judge Mann, his black robe hanging neatly nearby, waited in his chambers which he had supplied with ample pending paperwork as well as a large bag of his favorite red licorice.

With the judge's permission, Tom Fabel, Mike Cable, and the agents gathered at Mike's house fifteen miles away for a few hours and a few beers.

Susan stopped by during the evening to thank them, whatever the final outcome. She felt a little uncomfortable among them on their turf, so she left after a polite period, and the men didn't try to stop her. Susan was feeling more comfortable with herself, though, those days. As the psychologists often tell relatives of crime victims after a violent death, the trial can often take the form of a drawn-out funeral. It was hard to face all those sad facts every day, to poke at the scars. But Susan and the others told themselves that the real end was near when they could go on with life, scars and all. Eventually, the storm would pass. The tears would stop. The swollen eyes would subside.

Come August, on their wedding anniversary, Rudy's ashes would be buried in a small town in rural New

York State. He did end up in the country after all.
Despite her initial fears, Susan was running the bank
successfully without her husband. She commuted daily
from a rented house in Pipestone, where Rolph was
still in school. Her bank was for sale, and Lee Bush
would handle that. Some offers fell through, but the
final one was firming up. In early June she would be
free of those loan payments and that responsibility,
although for sentimental reasons Susan would deposit
a substantial part of Rudy's life insurance payments in
the Buffalo Ridge State Bank as a sign of support for
the town and her husband's dream. "Rudy would have
wanted that," Susan would confide, "and you know,
surprisingly, I did too." Surprisingly too for Susan, the
people of Ruthton were going out of their way to be
nice and helpful and friendly, Susan having now shared
some of the rural adversities like sudden death.

Life, and death, went on there as they always had.
Ruthton would lose again this year to Lake Benton in
the Homecoming game. A few babies would be born,
Greg Bot would commit suicide, and Wayne Onken
would be killed on the highway. In a while, Sharon
Fadness and her family would move away. But the
townspeople's expressions of sympathy to Susan and
appreciation for Rudy's contributions seemed genuine.
The Girl Scout troop even planted a tree in memory of
the man who financed their train trip to Duluth.

By early July, Susan had finally sold the Dallas
house and repaid Rudy's father. She decided to stay in
the Midwest. She bought her own modest bungalow
on a shady street in Minneapolis, where Rolph would
attend a private school, and she could do Junior League
work, and someday, who knows, maybe even go out
on a date again. She hadn't anticipated the tears,
though, that flowed when her furniture and belongings
arrived from storage and she began to unpack the
familiar but temporarily forgotten artifacts of a joint
life that ended so suddenly. She even found a box of
rocks that Rudy had collected in his childhood and
had carried everywhere since. Giving the Salvation
Army Rudy's clothes and shoes, including the wing

tips he wore That Day, was easy. Dumping out a box of anonymous sacred rocks proved harder.

Susan found herself—actually, her friends noticed it first—with a different attitude toward life, certainly a different shape. She had lost her extra weight. "I lived a pretty sheltered life," Susan said. "I was Betty Boop all my life, so much so that I didn't see and hear a lot of things and a lot of people." She was more religious after the shootings and didn't rail so much against the fates and fearful forces. "What choice do you have?" she would say. "You deal with the cards you are dealt." Susan also found herself wandering back out to the Jenkins farmhouse a few times, part of a migration route she developed to revisit familiar family touchstones. Perhaps it was just to make sure they were there before going on to new places. At the Jenkins place Susan walked along the same farm paths where she, Rudy, Rolph, and the dog had strolled and run. Susan always ended up out front in The Ditch, where she would stand and pray a little and mentally report to Rudy on what she was doing and thinking. She felt humbled there beneath that immense sky, standing where the lightning had struck and looking out across the flat fields that dwarf the humans who scurry across their surface. It felt good now.

The Jenkins house would soon be sold, for a song, to a family with children. So there would be life flowing back through its halls, and flowers painted on the mailbox in that ditch.

Getting Rolph through this time was more difficult. For a long while he desperately clung to Susan. In the afternoons after school he knew when she left the bank. He knew how long the drive took. If his mother wasn't walking in the door at the proper moment because she stopped to buy milk or cigarettes, Rolph was frantic, and angry, and frightened. If something unforeseen could snatch away one parent, then it could happen to the other. There was a tough time, too, in court that day when Rolph had felt like crying as Swen questioned Susan. The links between mother and son were very strong, if unspoken. Neither one would give that man the satisfaction of their tears. Until later, alone.

There were bad dreams and bad thoughts, too, and some guilt for Rolph. Susan found him sobbing once. The eleven-year-old remembered his father That Day after breakfast. Rolph went over and hugged him real tight before the man left for work. "And I'm always thinking," Rolph said, "maybe if I had held on to him longer and tighter, maybe he wouldn't have died."

At first, Rolph thought he might like to be a policeman and catch bad guys or maybe a prosecutor and do the same. A year later he wasn't so sure. A few months after that he wrote a theme titled "My Adventure." It was about the two-week canoe trip he and his father had taken in northern Minnesota that final summer. The theme described a violent storm that descended on their canoe in mid-lake, the wind, the whitecaps. But they got the tent up on an island, changed their clothes inside, played checkers, and read books. "All of a sudden," Rolph wrote, "we couldn't hear the rain anymore. We stuck our heads out of our tent and saw a beautiful sunset on the horizon."

Rudy's parents, still retired in Florida, often thought the pain of losing a son was past. "You expect parents to go first," said Rudy's father. Then his wife would be reading or painting and would see her husband sitting in a funny position and when she checked, the old man would be crying again.

Toby's mother still had the radio he brought home from the war. She wouldn't part with that for anything. She still lived with her physically handicapped son and she still taught music, although she couldn't take on as many students as before.

Toby's father was getting old too. His legs and stomach seemed to pain him more after September 29, 1983. He didn't go to the trial; he was afraid what he might say or do to the defendant.

"I have a firm belief in God Almighty," he said. "They say there's a reason for everything and I'm trying to find the reason, to see what could be the reason for this. As yet, I can't. Maybe someday. I don't know."

"No matter how you cut it," said Larry Thulin, "as a parent you have a favorite child. And he was it." Mr. Thulin left a light burning by his son's photograph at night where his second wife would find him sitting sometimes.

Mr. Thulin and his daughter-in-law Lynnette drifted apart after the death, though they were never too close before. Lynnette felt Toby's father sided with the other woman, Karen Rider. Toby's father thought his son was well rid of Lynnette.

Lynnette continued her audiology courses after Toby's burial, the only time Toby's two women ever saw each other. They were only brief glimpses really; friends kept the two females separated at the funeral home. Lynnette and Lynnette's children had a range of feature-length nightmares, as did Karen and Karen's.

Not long after the trial, Lynnette's father died; the lung cancer won out. Mother and daughter shared tips on coping with widowhood. As the months passed, Lynnette began learning from friends about a different Toby. "I knew he was no saint," she said. "He was human. He had a lot of problems, but we loved him." Like Susan, she would feel sorry sometimes for Steve Jenkins. Doomed from the beginning, they decided.

Sometimes Lynnette, who got her ears pierced and began wearing a little makeup, would think about all the problems she had encountered and feel sorry for herself. "Time helps me sit back," she said, "and laugh and cry and sort out the garbage and throw it away." She credits religion with carrying her through many difficult times. Religion and Toby. "He was my first love," she said. "There's only one of those in a woman's life. I'd do it all over again, even knowing what would happen."

Karen would not let go of her Toby either. "Ours was a mature love," she said. "I may have another relationship in my life, but there won't be another like Toby." Karen stayed in Minnesota for the trial, although it was suggested she avoid the courtroom. It was very difficult. Every day she'd see places and people and things that reminded her constantly of

Toby or, worse yet, the Jenkins family. There were some letters in the local papers, too, sympathetic to the poor Jenkins kid. Thursdays were the worst times for Karen, like sad anniversaries every single week, until one week over coffee Susan told her, "You can't throw away one seventh of your life. You've just got to get on with living."

Karen's dreams centered on rushing a gravely injured Toby to a hospital while a hating Lynnette stood watching every move. Karen wrote Lynnette some letters, which went unmailed. She wrote Toby's daughters, too, affirming how much their father loved them. Karen figured the Thulins didn't want to see her, so she marked some of Toby's possessions to go to them in the event of her death, something she thought more about after 1983. Karen's two children were growing up quickly, scarred from their losses, but good kids nonetheless.

Their nightmares seemed to dwindle after the three of them left Minnesota to settle in a house trailer in a small Colorado town where life would be peaceful and Karen could work more with handicapped youngsters. Even in Colorado, Toby's presence would pop up. A pack of his cigarettes in an old jacket pocket. A pile of his *Soldier of Fortune* magazines in a box. His familiar smell on a sleeping bag. And, of course, every Memorial Day Karen would scrape together enough money to wire a wreath to Toby's grave back home in Litchfield, Minnesota. Every year the inscription said the same: "To live in the hearts you leave behind is not to die."

There were no regular wreaths on Jim Jenkins's grave outside Paducah, Texas. It didn't take long for that sandy soil to settle and that long pile of dirt, the telltale sign of new death, to disappear into itself. There was some talk there about Minnesota and Ivanhoe when the trial witnesses from Texas returned home. Nice place, but they'd take home anytime. Sheriff Taylor was still making his regular rounds, still searching for an elusive vacation. Ted Beard and Jim Perry were back at their jobs and so was Charles Snow, who finally finished up the cattle shed and feeding pen that the two Jenkins men started building out on his ranch.

The old butane John Deere tractor that Jim Jenkins fixed up was still parked outside the maintenance shed. It was still available for any maintenance worker to borrow. But the big rubber tires were cracked and flat again, and no one seemed able to get the engine working the way Jim had. Not many tried.

The school that the Minnesota farmer guarded and helped to gut and then rebuilt with so much overtime was fully renovated. The termites were gone. Different critters, twelve-year-old boys in colorful jackets proclaiming allegiance to certain football teams, filled the halls and spilled down the front stairs, pushing and tripping each other and, just that year, beginning to pay attention to the other half of the school's student body.

Up the street Jackie Foster still opened the Kountry Kitchen at six and still served the same hamburgers and scrambled eggs. That day in April 1984 she would scan the Texas papers for news of a murder trial in Minnesota.

The jurors up in Minnesota, where the snowdrifts were almost gone and the tree buds almost there, pushed their dinner plates aside to continue their discussions, which were getting a little heated. There was no more sewing or card-playing now. For twelve people who had sat so silently for so long, they suddenly had a lot to say. Perhaps because they had sat so silently so long.

The waiting continued outside. Sheriff Abe Thompson was relaxing with his predecessor and present deputy, Vernon "Seeds" Dahl ("Dahl's the name, Law's the game"). To avoid every law officer's standard nightmare, a repeat of the Jack Ruby incident in Dallas, they had already made their contingency plans for a guilty verdict—get the kid out of the courthouse as quickly as possible. If Steve was found innocent, they wouldn't need any plans; the kid could walk out of there himself.

Abe put up a dollar that a verdict would come in

before midnight. Seeds put up two that it wouldn't. It was a continuation of a game the two had played for years, ever since Abe was a rookie policeman. Seeds would walk out of his office to bet the new guy twenty-five cents that he knew who had committed a particular crime. The new police officer knew there hadn't been any charges placed yet in that case, so he'd take the bet. The big, broad sheriff with the powerful grip that speaks louder than words would return in a few minutes with the signed confession that had been sitting on his desk while he went to file the charges. "That guy," said the sheriff, pointing at Dahl, "is as crooked as a corkscrew when it comes to betting."

The waiting continued throughout the evening as George Briffett, the janitor, went about his accustomed custodial rounds with unaccustomed company. Shifting groups of men and women sat, stood, talked, read, dozed. Some were so bored they began describing recent TV shows. A few strolled outdoors in the warming spring evening. Some crickets were back already from wherever they went in October. Some of those people waiting were getting paid overtime for their efforts. Others reaped a different kind of psychic satisfaction.

No one had expected an immediate verdict; there were, after all, six complex charges. But the absence of a decision after eight hours caused some concern. Would it be a hung jury and they'd have to do everything over again? Perhaps the delay meant imminent acquittal and one or two jurors were holding out for a conviction. Perhaps it meant one or two jurors were holding out for acquittal amid a majority for conviction. Maybe they'd decided quickly on one charge but were stuck on another. Or vice versa. Or maybe they hadn't decided on anything. The hallway debaters couldn't decide on what the jury hadn't decided on. Whatever it was, the delay indicated considerable wrangling.

"Now, I'm not easily scared," muttered Sheriff Thompson, looking over at the accused teen-ager, "but if I was him, I think I'd be quaking in my boots right about now." Steve was wearing boots, all right. He didn't appear to be quaking, although he had started to smoke.

At 11:48 P.M. a flurry of activity erupted. The prosecution team returned to the building, not so fresh from their private party, although their neckties were back on. The defendant's group made its way up the long stairs. The courtroom filled. The jury entered. And so did the judge. But it was a false alarm. Judge Mann wanted to tell the jurors that they could continue to meet until 1:00 A.M. or recess any time before then and resume in the morning. They should talk it over back in their room. They did. Recess time was set at 12:01 after the now ritual warning from the judge on talking about the case with anyone. He apologized for not having any overnight facilities, but he would have to rely on their integrity. They should return at 9:00 A.M. sharp but not restart their discussions until all twelve members were present, which was no problem since they all came in one or two car pools.

Everyone ran into storms on the way home that morning. Thunderstorms rumbled all around Lincoln County, the warm falling water helping to thaw the ground. Tom Fabel was feeling a little flushed from the evening's beers as he rode toward his distant motel to the rhythmic beat of the windshield wipers. Every minute or so his tired face was bleached an unnatural white by broad beams of lightning. His face might have been brighter had he known then that he was just three months and one day away from becoming the father of a very healthy—and very vocal—Theodore Thomas Fabel.

During a strenuous court case like Jenkins's Tom focuses on the accused. Rarely, Tom finds, is the trial their first brush with adversity, tragedy, or wrong. However, after his trial work is done, a tired Tom focuses on the victims, usually ordinary individuals plucked from the crowd by fate to suffer for reasons they will never know, if indeed there are any. At first they feel anger, outrage, a desire for revenge. Confusion and some anger often come with the methodical legal process too. "But when I get up close to these human tragedies," Tom says, "the resiliency of the human personality always impresses me."

These people have never seen themselves as heroic individuals innately full of courage and stamina to withstand the bad times until the sunshine returns. "They wear no badge of courage," Fabel says. They just do what comes naturally, surviving in their ordinary life, with varying degrees of grace. They still work. They still comfort their children, though they perhaps need more comfort themselves right then. As time passes and the wounds heal, they start to think of others, to smile more and to cry less. Years later, having survived and forgotten, they may see someone else's tragedy and someone else's successful struggle for survival, and they will say, "I don't know how they do it." And then they will go home and eat and go to bed, feeling very ordinary.

Swen Anderson hardly slept that night. He had farther to drive home, farther to drive back, and, it seemed, more at stake.

There was no formal court session at 9:00 A.M. on April 26. Everyone just fell automatically into his or her assigned place and assigned role of deliberating behind a barrier or waiting in the audience, as if the curtain were about to rise on the final scene of a play still being written.

Within minutes, the judge received a written note from the jury foreman: "We would like the definition of unconsidered or rash impulse. Could it be given in writing?" This referred to the judge's instructions in which he said an unconsidered or rash impulse, even though intended to kill, is not premeditated.

To Tom, this was good news. It meant the jury had decided Steve committed murder and was deciding now if he planned it in advance. To Swen, this was good news too. It meant the jury was having some real problems deciding on guilt. The reasonable doubt argument was reasonable after all.

Judge Mann wanted to know how the attorneys felt about providing the definitions. Swen said that was fine, but he'd prefer not doing it in open court as his client was going through an emotional wringer and one more appearance was unnecessary. The judge could just write out the definitions and send them back in. Tom said he'd always seen it done orally in open court, but whatever the judge wanted was okay with him. The judge had the definitions typed out from a *Webster's New Collegiate Dictionary*. The attorneys read them. The papers were sent to the jury. The wait continued.

The normal business of the courthouse continued too, with deeds being registered, property taxes being paid, and certificates for marriages, divorces, births, and deaths being filed. County officials were increasingly worried, however. They estimated the trial would cost more than forty thousand dollars, or nearly six dollars for every living person in the county.

Then at 11:19 word flashed out from the third floor: "There's a verdict!" Excitement swept the building. Most courthouse work stopped. Cars from around town converged on the courthouse. The spring sunshine was bright, though a tornado warning was in effect. Reporters ran for their regular seats. Courtroom artists sketched furiously. Camera crewmen donned battery packs and made last-minute checks. Susan and Sharon took their places. Lynnette was across the aisle. Darlene was nearer the front. The courtroom was full. Tom Fabel and Mike Cable calmly walked to their seats; they carried few files.

At 11:25 the jury walked in silently, looking carefully where they stepped lest someone stumble beneath so many stares. At 11:28 Swen entered with Steve, who was wearing blue jeans, a blue shirt, beige jacket, and his boots. As usual, he looked very calm and cool.

At 11:29 the judge entered, his black robe billowing out behind him. A serious Susan showed no emotion, though her knuckles did. Lynnette was praying to herself. Swen reached for Steve's hand.

"Ladies and Gentlemen of the Jury," said the judge, "have you reached a verdict?"

Dave Koster rose reluctantly. "We have, Your Honor."

The pharmacist handed the verdicts to the bailiff,

who carried them to Judge Mann. He opened them and read them silently while the courtroom waited. There were three pieces of paper—two verdicts plus a note from Dave Koster. If it's all the same to His Honor, the note said, could somebody other than the foreman read the verdicts out loud? The judge made no mention of the note but handed the verdicts to the court clerk, Lee Smith, to read.

"State of Minnesota. County of Lincoln. In District Court. Fifth Judicial District. Number K-83-254. State of Minnesota versus Steven Todd Jenkins, Defendant. Verdict of Guilty."

A gasp swept the courtroom. Lynnette's pink Kleenex was crumpled.

"We the jury find the defendant guilty of the charge of murder in the second degree as charged in Count Four of the indictment. Deems A. Thulin, victim."

The clerk picked up the next slip of paper.

"State of Minnesota. County of Lincoln. In District Court. Fifth Judicial District. Number K-83-254. State of Minnesota versus Steven Todd Jenkins, Defendant. Verdict of Guilty."

A louder murmur. Susan stared straight ahead.

"We the jury find the defendant guilty of the charge of murder in the first degree as charged in Count One of the indictment. Rudolph H. Blythe, Jr., victim.

"Is this your true and correct verdict as read, Mr. Foreman?"

"Yes, it is."

At Swen's request, Judge Mann then had the clerk poll each member of the jury on each verdict. While this long process was under way, Lynnette's tissue was getting wetter, Susan was staring straight ahead, Steven was still looking down, and Dave Koster was thinking. It had been an extremely arduous process, the pressure eating away at each juror every minute, and their emotions reflected that. It's tough being a leader in the countryside where a man can get his back up right quick if he's told what to do too bluntly. Dave had wanted a verdict vote right away. There was some debate, but he got it. He hadn't pushed too hard, more like a firm suggestion. The secret ballots were unanimous: Steve Jenkins had been involved in the murders somehow. Exactly how much, well, that would come later.

Did he do the shooting? A vote. Ten said yes, two said no. The discussion, which was to become a debate, began. They pored over their notes and memories, mentioning what struck them, where they noted discrepancies, what they believed, and who they didn't. The jurors never even discussed Darlene's testimony. They decided right off she had been protecting her son. They were very impressed by Fabel's presentation, such simple organization. The jurors hadn't realized how intelligent they were and how much they knew already about important things in life, about common sense, and even about the law. Several jurors commented on Fabel's dramatic closing argument, including the attorney throwing himself on the floor to aim the murder weapon. And the father's poor eyesight. And all that kid's training.

And some of the discrepancies. Who lied? Who would benefit by lying? For instance, who was driving when they left the farmyard? Why would a part-time policeman and the other guy lie about who was driving? Did Steve put himself in his father's position and say everything that actually happened as if he were Jim Jenkins? If you're not telling the truth, the easiest way to avoid getting caught is to put yourself in someone else's exact position, you know.

The two dissenting jurors slowly came around; they didn't have much to argue their case with anyway. All right, it was decided: Steve had done the shooting. But did he plan it?

Maybe Jim Jenkins and Blythe just got into an argument out there, some pushing and shoving, namecalling maybe. You know what bankers can be like. Jenkins, too. And the kid just opens up to protect his dad. Sure the kid talked about bulletproof glass and making fertilizer bombs, but he's accused of shooting, not bombing. A vote. It was nine to three that Rudy's murder was premeditated.

Wait a minute. Why take the license plate off if you're going to kill the guy? Well, he didn't have to plan it for months to be premeditated. What do you mean? He could have planned it for just a minute. That's premeditation, Fabel said. No, it was—what did the judge say?—a rash act. No. Let's get a definition from the judge.

When the judge's dictionary definition was delivered that morning, the unanimous verdict came quickly, which is not to say it made all twelve feel good. None would want to do that jury stuff again, though they didn't think of the retribution angle until later during the car ride home. There, maybe a neighbor who hadn't been at the trial said the kid was obviously innocent, and the juror discovered at supper that his or her child had taken some heat at school for a parent's decision.

The jurors were comfortable with the decision but worried about its impact. Some saw a kid's life going down the tubes. They didn't like feeling responsible, until someone noted they hadn't sent the kid out there to do the shooting. The foreman had said he was worried about choking up while reading the verdicts. Maybe he'd ask the judge to have someone else read them.

"Marcella Rieth, is this your true and correct verdict as read, murder in the first degree, Rudolph Blythe?"

"Yes."

That concluded the polling of the jury. The judge said a noon meal had been ordered for the jury. They could leave the building now or stay and eat, but their job was done. They were excused. All they told reporters outside was, "It was a very, very hard decision."

An appeal, sometimes taking years, is automatic in capital crimes in Minnesota: Steve's wouldn't come for nearly seventeen months. The defense—Swen would give way to a public defender—would claim there was no evidence of premeditation, that the judge should have sequestered the jury (though Swen didn't seek it himself), that the presence of a sheriff's deputy behind the lawyer's table was intimidating, and that the exclusion of his psychiatric testimony made a proper defense impossible. The Minnesota Supreme Court would disagree, however, upholding the verdicts in December 1985.

On that first verdict day the judge set May 22 as a sentencing date. At that time he would speak at length about the tragedy of the deaths and the tragedy of Steve's life. He would criticize a certain military officer, whose skills were honed in the duty of his country training young men to be soldiers, for teaching yet another teen-ager some of the mechanics of firing a deadly weapon accurately. The sentencing was a formality, since first-degree murder carries a mandatory term of life, which in Minnesota means a minimum of seventeen and a half years. Judge Mann, impressed with Steve's new family life with the Andersons, would rule that the ten-year sentence for Toby Thulin's murder and the five years for firing on Deputy McClure's car could be served concurrently.

Given the seriousness of the crime, however, Judge Mann revoked bail at the end of the trial. The teenage defendant Steven Todd Jenkins was to be turned over to law enforcement authorities for transportation to begin his imprisonment at least until the year 2001, when he would be thirty-six.

"Court's adjourned."

Word of the verdicts swept through town and out into the countryside and even down into Texas long before the evening news put the pictures in the air.

Everyone stood up as the judge left. Steve was calm, though noise and people and pandemonium began crowding around him. He remained calm until Seeds Dahl, the big beefy deputy, walked toward him, according to Sheriff Thompson's plan. Dahl reached around his back to his belt and pulled out a pair of chrome handcuffs like the ones Steve used to carry.

Steve broke down then, sobbing and clinging to his mother. Darlene Jenkins and Steve stood there hugging and crying for several minutes while dozens watched.

Four feet away, reporters—under sudden competitive deadlines now that something had happened—prodded Fabel for a reaction. "There is no sense of elation or victory, no sense of vengeance," he said. "It is the sad culmination of a tragedy for three separate families that began last September twenty-ninth, and we can only hope now that the families can begin to put their lives back together again."

Deputy Dahl put his large right hand on the back of Steve's neck. "Okay, boy," he said, and a crying Steve was gently pried from Darlene. They went out the courtroom door, the three of them together, Darlene hanging on Steve's arm, and they moved down the long metal stairs.

At the bottom Swen was expressing his love for the boy to the television cameras and vowing a thorough appeal. But his last few words were ignored as all the cameras and bright lights swung toward the newly convicted murderer.

"Seeds, will you wait for me?"

"If you hurry, Swen."

The mob grew as the deputy and Steve burst out the heavy courthouse door into the sun. Cameramen connected to their crewmen by black wires tried to run backward in unison through the pushing crowd, attempting to get a head-on shot of the youth and the officer. Microphones on long poles waved in faces. Seeds was not taking his time out here in the open. Down the sidewalk, around to the right, back toward the sheriff's garage and a waiting patrol car. Steve was put in the back with Dahl. Swen got in the front. Doug Pedersen, another deputy, drove. At 11:51 the police car moved off toward the jail. It had been parked precisely on the same spot where authorities had put the bloody Blythe vehicle That Day.

Out front, Tom Fabel had finished speaking into a forest of microphones. "There is no sense of elation or victory . . ." Back upstairs packing his briefcase, the prosecutor had other thoughts. "I'm awfully sad about the human condition that brings these things to pass," he said, "but I'm going to enjoy going home now and putting my arms around my wife and kids and reminding myself that life is not all death and sadness."

In the hall George Briffett was leaning on his big broom, surveying the afternoon's work ahead. The refuse included cigarette butts, coffee cups, soda cans, potato chip bags, candy wrappers, newspapers, dirty dishes, and a crumpled pink Kleenex. "There's so much to clean up," he said.

All that while, Lynnette Thulin sat alone on a bench outside the courtroom. Her eyes were swollen. But the tears had stopped.

"Amen," she said.